GAME THEORY

A NON-TECHNICAL APPRAISAL OF GAME THEORY WITH NEW DIMENSION

SAMARJEET TRIPATHI

ISBN 978-1-68586-315-9

Dedicated to

MAA AADISHAKTI SHARDEY & SHARDEY SADAN FAMILY

AND MY LATE DADA JI

Contents

Acknowledgements

The completion of this undertaking could not have possible without the participation and assistance of so many people whose names may not all the numerated.their contributions are sincerely appreciated.

This book is the brainchild to establish game theory as a seperate subject. GAEORY as a subject would be universal and subject of next generation. it has the distinction of intimate interaction between theory and practice.

writing this book has an exercise in sustained suffering and sacrifices. it is worth as them that i love, know that i love them. I want to thank my readers and academicians.

this is my first academic book, i would like to thank to my destiny,failures,pains and sacrifices. special thanks are due to those who gave their time to read or edit my manuscript.

Thank you, divine spirit mystery of universe and nature.

Gratitude

Hello homo!

Are you in dilemma- personal or professional one?

Confused about taking decisions of education, business or personal?

Want to start your Start up?

Want to enter in politics or administration?

Or whatever the case be- this is right place to get basic idea about game theory, which has all solutions to every problem!

Come on – have look at this, it is as simple as driving car; but hold on, ride a bicycle first, good for health and climate as a whole, clear... global warming already above the very high to pre industrial era

It`s is no more just a subject of economics and technical – with new avatar, GAEORY is for everyone

hey you...

backbencher, be in Personal or Professional, Rikshaw puller, Office boy and Art students and of course, and housewives,

hello ladies,

don't disappear through the crack

come on, you are very good at balancing the workforce and households

hello dear India,

your women bear the burden of unpaid work, let empowered her

hello world

lot of cracks you are in like continental drift theory, come on fixing it

GAEORY is subject of next century – Samar J. Tripathi

Preface

I tried to write the book in such a way that it would be accessible to anybody with minimum knowledge of mathematics and no prior knowledge of game theory.

However, my upcoming book in GAEORY would intend to be rigorous and it will include several proofs and technical analysis.

Despite my best effort, there will be certainly some errors or spelling mistakes for which I alone responsible.

I would be grateful if you could reached at; 10samar.1994@gmail.comwith suggestions for making improvements in the book from fellow scholars are most welcome.

I am very grateful to **Dr. Nandita Agrawal**(professor at university of Allahabad, ADC); for meticulously going through the concept of game theory in university lecture and for suggesting numerous improvements in my methodology. *Dr. Mukta Tripathi and Dr.Deepshikha for their, unwavering support and* encouragement, as well as their always insightful and constructive remarks. Have significantly improved this work.

I would also like to thank my family specially **parents**and **brothers– Mr. *Indra jeet,*** **Mr. Satya jeet**and my mentor PCS Mr. Dharmjeet.

With TCR (vasudhaiva kutumbakam) Philosophy, my dear Father – you shaped us into perfect human. MAA - you are our world.

you have shielded me like an ultimate saviour from infancy, and your steadfast support has always served as a boost to my intellectual index; you are our pride and love glue that hold our brothers together – Mr. Dharmjeet : thank you for your kindness and admiration.

I`d like to express my gratitude to my friends Shivani Rathod, Anirudh B., **Priyanka ji**(Madhya Bharat, weekly newspaper), Kumaresh mandal (presidency college, mathematics dept), Mr. Hari sharan, Mr. Alwin, Er. Abhishek, Drx Amarjeet Tripathi, Drx. Ashutosh and humanity as whole.

I must say thank you to Mr. Gyan Prakash, Vaishnavi Tripathi, Saraswati Tripathi, Srishti Mishra and my brilliant Sister-in-laws Mrs. Arti; Mrs. Swati and Mrs. Neha tripathi for pointing out typos and helps.

Foreword

Start with a statement from Dan Brown: "Life is full of difficult decisions, and the winners are those who make them". So this book will guide to everyone to take difficult decision with the help of GAEORY.

I was delighted when Mr. Samarjeet informed me about his book on this unique subject with new dimension, and I was anxious to learn more about it. It was an honor for me to be approached for recommendations for his book. Coming from a computer science background, economics and social studies are not my strong suits, which is why I conducted some research into Game Theory. Undoubtedly , author way of putting his theory in non-technical way and generalization of theory is commendable.

The majority of the book is focused on **"Game Theory as a subject (GAEORY),"**which looks at individual decision-making scenarios through a number of scenarios. The author was directly involved in evaluating numerous situations and components of the game theory, and this book pays special tribute to economists who worked on GAEORY research and analysis. This book details a person's strategies and planning as they move through the decision-making process, as well as their current mental state.

Every day, we witness many outstanding people, whether they are CEOs of big businesses, leaders of countries or governments, or even ourselves, making critical and tough decisions. Even if someone else makes a decision on your behalf, you have the last say on whether or not that person has the authority to do so. That is why individuals must understand how important it is and how to put it into action with all of the required planning and strategy in order to make the most rational and reasonable decision possible.

And, for writing this book, I wish **Mr. Samarjeet Tripathi**, the author and one of my closest friends, every success with his book on "Game Theory" and upcoming "GAEORY."

Neha choudhary

(NEHA CHOUDHARY)

Abstract

"Find a truly original idea. It is the only way I will ever distinguish myself. It is the only way I will ever matter." – Nobel laureate John Nash"

Game Theory is a field of study that help us understand decision making in strategic situation, in addition to being an important methodology within the economic discipline, it also gives insights into pricing and management strategies used by business. Furthermore, game theory has wide - ranging applications in areas such as, international relation, political science and military strategy. Much of game of theory involves the interaction of decision makers where there is an asymmetry of information. Thus, the study of game theory can provide insights into how decision makers act when there is some important information that they cannot directly observe.

While giving overview to critical appraisal of game theory, the application of prisoner`s dilemma even discussed, before giving the shed light on GAEORY as a subject.

Prisoner's dilemma game is a deceptively simple 2*2 matrix game which can be used to illustrate the value and limitations of game theoretic thinking. Its simplicity makes it most attractive as a paradigm to explain human behaviour. Analogies between it and human affairs are best employed to study their inadequacies and to pinpoint what has been left out rather than to claim how much of the world can be packed into 2*2 matrix.

Game theory has had a profound influence on many fields of the social sciences since its rise to prominence more than eighty years ago. The game theory presents a technical analysis of strategic interactions. So far, we have focused on its applications and are very vast. But as per my non-technical analysis, game theory is the subject of next century.

Game theory as a subject"GAEORY", the word which I, propounded in my analysis to accelerate this as an independent discipline and not just borrowed concept.

UNIT - I

"The entire edifice of game theory rests on two theorems: Von Neumann`s min-max theorem of 1928 and Nash`s equilibrium theorem of 1950 ... Nash introduced the distinction between cooperative and non-cooperative games ... By broadening the concept to include games that involved a mix of cooperation and competition, Nash succeeded in opening the door in applications of game theory to economics, political science, sociology, and, ultimately, evolutionary biology." (Math & Genius & Games & Economics & Politics & Evolution) ibid.

The Game theory is the study of decision making under competition. More specifically, Game theory is the study of optimal decision making under competition when one individual`s decision affect the outcome of a situation for all other individuals involved.

Game theory better described as 'Interactive Decision Theory'. The first studies of games in the economics literature were the papers by Cournot (1838), Bertrand (1883), and Edgeworth (1897) on oligopoly pricing and production. Borel then gave the first reasonably systematic treatment of game theory in 1921, although he did not provide any rigorous proofs for his speculations. This was built upon by von Neumann who went on to prove a central concept – the minimax theorem- in 1928, before collaborating with Morgenstern to publish the Theory of Games in 1944 (Leonard 1992).

In paper, Fun, Games & Economics – James Jerome Lim mentioned that; despite the initial excitement following the publication, game theory spent a long period in doldrums, as economists were slow to see the importance of the theory. Game theory was pigeonholed as a theory for the small numbers case in economics, and its popularity waned (Schotter & Schwodiauer 1980).

However, in the 1950s game theory enjoyed a revival for three reasons: first, the ascendancy of mathematical economics; second, the widening of the field of game theory as it was applied to the analysis of general equilibria; and third, the strong interest of the military (Leonard

1992;Schotter & Schwodiauer1980).

From then on, game theory grew from strength to strength. Its continued growth in terms of both theoretical extensions as well as widened applications in other areas of economics proved that game theory theorists in 1944 sealed the intimate relationship between economics and game theory. But it also widened its scope to other related subjects as well.

Game theory is in its evolutionary stages. So, it would not to be a surprise to us to see that – game theory is the subject of next century i.e. **GAEORY.**

CHAPTER ONE

CHAPTER 1

Introduction

Games theory works in terms of self-interest. Some game theory concept could be unsound. That`s overdependence on rationality. That is my enlightenment.

-John Nash

Game theory is a theoretical framework for conceiving social situations among competing players. In some respect, game theory is the science of strategy, or at least the optimal decision-making of independent and competing actors in a strategic setting.

Game Theory is a set of analytical tools developed for understanding situations of **interaction** between (rational) individual decision makers (economic agents, social entities, participants to a game). The key to game theory is that one player payoff is contingent on the strategy implemented by the other player. The game identifies the players` identities, preferences, and available strategies and how these strategies affect the outcome. Depending on the model, various other requirements or assumption may be necessary.

The Prisoner's dilemma is a standard example of a game analysed in game theory that shows why two completely rational individuals might not cooperate, even if it appears that it is in their best interests to do so. Its relevancy applies almost every aspect of life; be it social, political, economic or so.

So, basically Game theory is a bag of analytical tools designed to help us understand the phenomena that we observe when decision-makers pursue well defined exogenous objectives (rational) and take into account their knowledge or expectations of other decision makers` behaviour (they reason strategically).

SECTION: ONE

Mystery solved? I think so.

Let`s first take a look at,"The critical appraisal of game theory & the relevance of prisoner's dilemma in oligopoly market"

The theory of games is one of the most outstanding recent development in economic theory. It was first presented by Neumann and Morgenstern in their classic work.

"Theory of Games and Economic Behaviour", 1944 has been regarded as a 'rare event' in the history of ideas.

Game Theory grew as an attempt to find the solution to the problem of duopoly, oligopoly and bilateral monopoly. In all these market situations, a determinate solution is difficult to arrive at due to conflicting interests and strategies of the individuals and organisations. The Theory of Games attempts to arrive at various equilibrium solutions based on the rational behaviour of the market participants under all conceivable situations. The ideas behind Game Theory have appeared through-out history.

The underlying idea behind game theory is that each participant in a game is confronted with a situation whose outcome depends not only on his own strategies but also upon the strategies of his opponent. It is always so in chess or poker game, military battles and economic markets.

Thus, Game Theory is a mathematical concept, which deals with the formulation of the correct strategy that will enable an individual or (entity) i.e., player, when confronted by a complex challenge, to succeed in addressing that challenge.

Game Theory was developed based on the premise that for whatever circumstances, or for whatever game, there exists a strategy that will allow one player to win. Any business can be considered as a game played against competitors, or even against customers. Economists have long used it as a tool for examining the actions or economic agents such as Firms in a market.

Thus, Game Theory is helpful in providing solutions to some of the complex economic problems even though as a mathematical technique, it is still in its developing stage, despite many limitations as well.

The game of prisoner's dilemma is of important relevance to the market structure specially oligopoly, as prisoner's dilemma is a game that explain why it may be hard to maintain cooperation of oligopolists at any time, including when there are mutual benefits. In this game, two criminals are usually arrested, and it is generally up to them to decide whether they will cooperate or they will rival against each other. The dilemma comes in that whereas the two prisoners' will benefit more by working together, each puts their own interests first. Just as in the prisoner's dilemma, it isn't easy to maintain cooperation in the oligopolies, this is because collaboration is not one of an individual's best interests.

For example, commodity A and B are competing against each other where each is produced by two different companies operating in an oligopolistic market.

Assume that both commodities A and B are types of soft drinks, and one of them changes in price.

If commodity A lower its prices, commodity B may be forced to follow suit to ensure they do not lose their clients. However, commodity B may choose not to lower its prices as commodity A, thus sacrificing its market share. Arguably, both companies would have benefited if they agreed on pricing strategies to use.

SECTION: TWO

What is Game Theory

It is the tool to solve all your problems

"Well, if the rules of the game force a bad strategy, maybe we shouldn`t try to change strategies. Maybe we should try to change the game"

- Brian Christian & Tom Griffins

Game Theory is a theoretical framework for conceiving social situations among competing players. In some respect, game theory is the science of strategy, or at least the optimal decision-making of independent and competing actors in a strategic setting.

They key pioneers of game theory were mathematician John Von Neumann and economist Oskar Morgenstern in the 1940s. Mathematician John Nash is regarded by many as providing the first significant extension of the Von Neumann and Morgenstern work.

Any time we have a situation with two or more players that involve known pay-outs or quantifiable consequences, we can use game theory to help determine the most likely outcomes. In simple term, Game Theory defined as the process of modelling the strategic interaction between two or more players in a situation containing set rules and outcomes.

Game theory has been widely applied to the behaviour of producers with a few or only one competitor.

Basically, all games have rules, which govern conduct of the players; pay-offs, such as win, lose or draw strategies, which influence the decision-making process.

In applying game theory to the behaviour of firms we can suggest that firms face a number of strategic choices which govern their abilities to achieve a desired pay-off, including:

Decisions on price and output, such as whether to:

- Raise
- Lower
- Hold

Decisions on products, such as whether to:

- Keep existing products
- Develop new one

Decisions on promoting products, such as whether to:

- Spend more on adverting
- Spend less
- Keep spending constant

Firms could derive a range of possible pay-offs from their strategy choices, including:

- More profits for shareholders
- Greater market share
- Improved chances of survival
- Getting rid of a rival

Before we start the analysis of the Theory of Games and relevancy of prisoner's dilemma, it will be useful to digress on certain fundamentals of game theory.

A game has set rules and procedures which two or more participants follow. A participant is called a player. A strategy is a particular application of the rules leading to specific result.

A move is made by one player leading to a situation having alternatives. A choice is the actual alternative chosen by a player, the result or outcomes of the strategy followed by each player in relation to the other is called is pay-OFF;

The saddle point in a game is the equilibrium point.

There are two types of games: constant-sum and non-constant-sum.

In a constant-sum game what one player gains, the other loses. The profits of the participants remain the same.

In a non-constant-sum game, profits of each player differ and they may cooperate with each other to increase their profits.

The Game theory can be divided into two main subdisciplines: classical game theory and combinatorial game theory.

i. Do you still remember your child-days play- rock, paper, scissors?
ii. Do you recall Anand, Ger renowned players of chess?

Take a deep breath and let`s come back to the topic that is classification of game theory.

a. *Classical game theory* studies games in which players move, bet, or strategize simultaneously. As a result, players often find themselves ignorant to certain aspects of the game. Players of these games are more likely to depend on prediction and chance due to this lack of information. Examples include poker or rock, paper, scissors.
b. *Combinatorial game theory*, on the other hand, is the study of two-player games in which each player has complete knowledge of all aspects of the game throughout the entirety of game play. These games are usually played on a turn-by-turn basis and do not typically involve elements of chance. Examples include chess or checkers. Furthermore, combinatorial games are said to be impartial if all players have the same set of possible moves from each position. Otherwise the game is said to be partisan.

SECTION: THREE

WHY GAMES THEORY

Because the God created universe & earth with many mystery

"I think game theory creates ideas that are dominant in solving and approaching conflict in general"

- Samarjeet Tripathi

Interaction between (rational) individual decision makers (economic agents, social entities, participants to a game).

Games that humans invent of course, hence the names but also many social activities, Military strategy making, Political decision making, also Types of network & Understanding human reactions to any policy or regulation strategy, Near all economic activities; competition between firms on a market, investment and research & development strategies of firms, inflation management by a central bank; But interactions are difficult to analyse using traditional tools of political science, sociology and economics:

What could/should the individual behaviours because in such situations?

What could be the collective outcome in this situation?

How to design the situation (games in order to obtain a desired outcome?

Necessity of building new tools: The game theory reframes such situations as will structured games that we can analyse.

Thus, Game theory offers us a great understanding of strategic interactions and corresponding decision making.

Generally, game theory can be used to achieve two goals:

- Undertake precise analysis of interaction in particular strategic settings, with a view to predicting the outcome
- Design rules of the game in such a way that self-interested agents behave in some desirable manner (i.e. tell the truth); this is called mechanism design;

Both are these approaches are quite useful for the study of argumentation in multiagent systems to choose the best strategy.

SECTION – FOUR

"We are living in a whole new social and economic order with a whole new set of problems and challenges. Old assumptions and old programs don`t work in this new society and the more we try to stretch them to make them fit, the more we will be seen as running away from what is reality. So dynamism even needed in economic model`s assumption" - ***Samar jeet tripathi***

ASSUMPTIONS OF GAME THEORY

Game theory is a study of strategic decision making. More formally, it is the study of mathematical models of conflict and cooperation between intelligent rational decision makers. Individuals those are instrumentally rational always have first choice over different things. Technically individual must have preference ordering that will make them to formulate judgement above various actions fulfil our preferences in diverse extent.

In game theory we usually make assumptions; For instance, the number of participants in the game must be finite. All participants are assumed to act rationally. Player should have detail knowledge about the game and his opponent. There also should be conflicting interests of the participants. Obviously, earth would be an uninteresting place to dwell in if it was entirely free of conflict!

The number of strategies available to each participant should be finite and may vary from participant to participant. here, a strategy refers to a complete plan of action a player will take, given the set of circumstances that might arise within the game.

Why is Game theory considered part of mathematics?

Game theory started as a branch of applied mathematics because its pioneers – Von Neumann, Nash and Shapley – were physicists and mathematicians. The idea of equilibrium in game theory is connected to the existence of fixed-points (e.g. Kakutani`s fixed point theorem). Strategic situations in economics settings that we are all familiar with (e.g. Prisoner`s dilemma) are simply useful and important applications of these mathematical results which consists logical arguments.

In my book, I tried to make game theory concept more simple & non-technical as much as possible, because my purpose is to teach the importance and application of game theory even to common people. I see game theory as a separate subject namely GAEORY.

CHAPTER TWO

Chapter 2

Index Ludorum

It will make easy the concept of game theory and my next book on GAEORY

A Game, more precisely, can be defined as a set of circumstances that has a result dependent on the actions of two or more decision makers (players). A move in this game is a point where the players are faced with choices.

Games can be classified into three classes:

games in normal form, games in extensive form and games in beyond normal and extensive form. Normal and extensive form games are further divided into zero-sum, non-zero-sum, perfect information, imperfect information games. Beyond normal and extensive form games are further divided into repeated, stochastic, Bayesian and congestion games.

SECTION: ONE

Classification of Games:

A game, in the mathematical sense, is a situation in which players make rational decision according to defined rules in an attempt to receive some sort of payoff. Game theory is the branch of mathematics which focuses on the analysis of such games.

Game theory can be divided into two main subdisciplines, namely classical game theory, in which players more, bet or strategize simultaneously, as a result, players often find themselves ignorant to certain aspects of the game; and combinational game theory, on the other hand, is the study of two-player games in which each player has complete knowledge of all aspects of the game throughout the entirety of gameplay.

In addition to the two classifications, games can be classified in a variety of ways. One of the most obvious is to classify a game by the number of players. It's common to describe a game as an n-person game, where n is an integer greater than or equal to 1 representing the number of players required to participate in a particular game.

The order in which players move (or lack thereof) is another easy way to classify game. Players all makes their move at the same time in simultaneous game. Contrarily, in a sequential game, only one player may move at any given time.

Games can also be classified based on the total possible winnings. A constant-sum game or zero-sum game is one in which the sum of total possible winnings remains constant no matter what actions players take, that is, the sum of the winnings gained by some players must be equal to the sum of the other players losses.

In poker, for example, players compete for a constant sum of money. The decision of each player does not affect the available winnings. In variable-sum games, however, the total available winnings may change depending on the player's actions. The prisoner's dilemma is an example of a variable-sum game.

Variable-sum game can be divided even further into following subgroups: cooperative and non-cooperative games. Players of cooperative games are permitted to make binding agreement, such as an enforceable contract, while players of non-cooperative games may not create any binding arrangements. For example, imagine there are two individuals, a seller and a buyer, hoping to complete a business transaction. As they attempt to negotiate a price, the individuals are participating in a non-cooperative game. if the buyer signs a contract agreeing to pay a specific price, it then becomes a cooperative game.

The difference between sequential and simultaneous games:

Simultaneous games are the one in which the movement of two players is simultaneous. In the simultaneous move, players do not have known about the move of other players. On the contrary, sequential games are the one in which players are aware of the moves of players who have already adopted a strategy.

However, in sequential games, the players do not have a deep knowledge about the strategies of other players. For example, a player has knowledge that the other player would not use a single strategy, but they are not sure about the number of strategies the other player may use. Simultaneous games are represented in normal form while sequential games are represented in extensive form.

In sequential, they know prior knowledge of the opponents m

moves but in simultaneous, they don't. Sequential is more of an extensive game. An extensive game is allowing explicit representation of a number of important aspects, like the sequencing of players' possible moves, their choices at every decision point, the information each player has about the other player's moves when he makes a decision, and his payoffs for all possible game outcomes. Whereas Simultaneous is more of a strategy game. A strategy game is in which the players' uncoerced, and often autonomous decision-making skills have a high significance in determining the outcome. Almost all strategy games require internal decision tree style thinking, and typically very high situational awareness.

Basically Games can be classified as –

- **Based on strategy -**

- Game of pure chance e.g. lotteries, slot, machines
- Games of strategy e.g. chess, Go
- Game of strategy and chance e.g. poker, monopoly

- **Based on cooperation**

- Cooperative game
- Non-cooperative game

- **Based on movement**

- Dynamic game also known as sequential game
- Static game also known as simultaneous game

Brainstorming concepts need to be revisited– few amongst are:

- A stronger solution concept in game theory is the dominant strategy equilibrium. this is a strategy profile where each agent is playing a dominant strategy. this is a very robust solution concept since it makes no assumptions about what information the agents have available to them, nor does it assume that all agents know that all other agents being rational (i.e. trying to maximize their own identity). – analyse this in the concept of government forming , where every legislature(party) has dominant strategy specially in hung parliament.

- The Bayes-Nash equilibrium
- The Revelation principle is a powerful tool when it comes to studying implementation. It determines whether a particular social function can be implemented. Rarely sees in the real world because,

a. it can place high computational charge as required to execute agents` strategy
b. agent strategies may be computationally difficult to determine, and
c. agents may not be willing to reveal their true types because of privacy concerns.

- What game rules guarantee a desirable social outcome when each self-interested agent selects the best strategy for itself? Read – Re-read...well pointed right, this is the case of *reverse game theory. This has its own importance in democratic setup, be it family, society, nation or global institution.* A sub field of game theory which concerns reverse game theory known as Mechanism design. So while game theory is concerned with a given strategic situation modelled as a game, mechanism design is concerned with designing the game itself. Don`t you think Argumentation mechanism design works as like Game-maker. That is where, GAEORY concept comes which has macro(broad) concept.

So, there are many other concepts, which would be analyse via theorems and applications in my upcoming book on GAEORY. So you have plenty of times to critically analyse game theory concept in non-technical way.

Thus, classification could be described in a more elaborative way, which would be suffice to correlate non-technical analysis approach of game theory.

Thus, critical appraisal of game theory and relevancy of prisoner's dilemma can be discussed pragmatically. This topic is vast so taking brief sum-up approach to present here before upcoming work on GAEORY.

SECTION- TWO

It is not easy to compete with yourself first, not others: TCR PHILOSOPHY

One Person Games

Because, Every vote matters

One person games are also known as games against nature, with no opponents, the player only needs to list available options and then choose the optimal outcome. When chance is involved, the game might seem to be more complicated, but in principle the decision is still relatively simple. For example, a person deciding whether to carry an umbrella weighs the costs and benefits of carrying or not carrying it. While this person may make the wrong decision, there does not exist a conscious opponent, that is, nature is presumed to be completely indifferent to the player's decision, and the person can base his decision on simple probabilities.

As we know, in addition to game theory, economic theory has three other main branches: decision theory, general equilibrium theory and mechanism design theory and all are closely connected to game theory. Decision theory can be viewed as a theory of one person games, or a game of single player against nature. In making decision, probability theory is heavily used in order to represent the uncertainty of outcomes, and bayes law is frequently used the model the way in which new information is used to rerise beliefs. Thus, one-person games, decision theory is often used in the form of decision analysis, which shows how best to acquire information before making a decision

SECTION- THREE

Only the fittest will survive - Darwin whispered in my ear: Samarjeet Tripathi

Zero-Sum game or Two- Person Constant-Sum game

Game theory provides a mathematical framework for analysing the decision-making processes and strategies of adversaries (or players) in different types of competitive situations. The simplest type of competitive situations is two-person, zero-sum game. These games involve only two players; they are called zero-sum games because one player wins whatever the other player loses.

Example: Odds and Evens

Take a simple game called odds and evens. Suppose that player1 takes evens and player 2 takes odds, then each player simultaneously shows either one finger or two fingers. If the number of fingers, then the result is even, and player 1 wins the best (? 2k). If the number of fingers does not match, then the result is odd, and player 2 wins the bet (? 2K). Each player has two possible strategies: Show one finger or show two fingers.

SECTION- FOUR

Data is new oil, so information is new energy

Games of Perfect Information

Fight for information, not for bread that's mantra for survival in ICT Era

"A proven theorem of game theory states that every game with complete information possesses a saddle point and therefore a solution."

-Samarjeet tripathi

A class of game in which players move alternately and each player is completely informed of previous moves.

Finite, zero-sum, two player games with perfect information (including checkers and chess) have a game saddle point, and therefore one or more optimal strategies.

In 1912 the German mathematician Ernst Zermelo proved that such games strictly determined; by making use of all available information, the players can deduce strategies that are optimal, which makes the outcomes preordained (strictly determined). In chess, for example, exactly one of three outcomes must occur if the players make optimal choice:

i. white wins (has a strategy that wins against any strategy of black)
ii. black wins or
iii. white and black draw.
iv. Thus, in game theory, a sequential game has perfect information. If each player, when making any decision, is perfectly informed of all the events that have previously occurred, including the "initialization event" of the game (e.g., the starting hands of each player in a card game).

Perfect information is importantly different from- complete information, which implies common knowledge of each player's utility functions, payoffs, strategies and types. A game with perfect information may or may not have complete information.

Thus, chess is an example of a game with perfect information as each player can see all the piece on the board at all times. Other examples include tic-tac-toc, checkers and go.

SECTION- FIVE

Access Denied – Oops, you are with imperfect information

Games of Imperfect Information

Only way to survive in ICT era – get perfect information at the right time

"There`s a rule that you really only want to play one level above your opponent, 'explains poker professional Vanessa Rousso. If you play to far above your opponent, you`re going to think they have information that you want them to glean from your actions. So better seek out games where honesty is the dominant strategy"

- Samarjeet Tripathi

Informally, a game of incomplete information is a game where the players do not have common knowledge of the game of being played.

Among the aspects of the game that the players might not have common knowledge of are :

- Payoffs
- Who the other players are
- What moves are possible
- How outcome depends on the action
- What opponents knows, and what he knows I know.

To take a couple of simple examples:

1. In price or quantity competition, firms might know their own costs, but not the costs of their rivals;
2. The government may design the tax code not envisioning what ploys people will come up with to avoid taxes;
3. Countries may negotiate climate change agreements having different beliefs about the costs and benefits of global climate change;
4. Plaintiffs may offer settlements to defendants not knowing what sort of case the defendant will be able bring to court, or what sort of case the defendant thinks the plaintiff will be able to bring.

5. Firm investing in R&D might know how their project is coming along, but have no idea who else is working on the same problem, example; in case of Covid-19 vaccine R&D case.

Games where some aspect of play is hidden from opponents such as the cards in poker and bridge- are examples of games with imperfect information.

So, a game of imperfect information is a game where the players do not have common knowledge of the game being played. Thus, imperfect information appears when decisions have to be made simultaneously, and players need to balance all possible outcomes when making a decision. A good example of imperfect information games is a card game where each player's card is hidden from the rest of the players.

***Hey**, don`t be bored; thora sa mathematical analysis Jaroori (little bit mathematical analysis and we need, to give respect to the mother subject of GAEORY i.e. Mathematics)*

So, an imperfect-information game is an extensive form game in which each agent's choice nodes are partitioned into information sets.

- As information set = all the nodes you might be at

The nodes in an information set are indistinguishable to the agent.

So, all have the same set of actions

A perfect-information game is a special case in which each I set contains just one node h

So, treating an imperfect-information game as a collection of perfect information games has a theoretical flaw as it can`t reason about information-gathering moves. In bridge, that didn`t cause much problem in practice, but it causes problems in games where there is more uncertainty: in such games, information-set search is a better approach.

As so, the paranoid opponent model works well in perfect-information games such as chess and checkers. But the hidden-move game that we tested, it was outperformed by the overconfident model. In these games, the opponent doesn`t have enough information to make the move that`s worst for you. It`s appropriate to assume the opponent will make mistakes

SECTION- SIX

Hey, don`t put all your eggs in one basket, diversification is key to the future – Samarjeet tripathi

Mixed Strategies and The Minimax Theorem

A strategy, which helps to maximizes the payoff among worst outcomes

In simple, Mixed strategies is a strategic game, when the player does not choose one defined action, but rather, chooses according to a probability distribution over a her actions.

Whereas, Minimax strategy means, a strategy of always minimizing the maximum possible loss which can result from a choice that a player makes.

When saddle points exist, the optimal strategies and outcomes can be easily determined. However, when there is no saddle point the calculation is more elaborate. So, in some cases, there is no saddle point, and players have to choose their strategies with a degree of randomness, as in the following simple game, called "Matching pennies". Two players simultaneously place a penny on a table, either heads up or tails up. If the pennies are the facing the same way, player1 gets to keep both pennies. Otherwise, player2 gets to keep both. So a mixed strategy Nash equilibrium involves at least one player playing a randomized strategy and no player being able to increase his or her expected payoff by playing an alternate strategy. A Nash equilibrium in which no player randomizes is called a pure strategy Nash equilibrium.

There is no clearly defined strategy for each player. The best way to play is to choose the position of the coin randomly. If either player follows this strategy, then in the long run, the payoffs for each will be 0. Notice that if, say, player1 uses a 50/50 strategy, while player2 plays heads 75% of the time, in the long run, both payers will still have payoffs of 0, but if player2 follows the 75/25 strategy, then player1 can easily take advantage of it by playing heads more frequently and therefore winning more frequently. So, it's important for each player to not only maintain a random strategy, but to also analyse the strategy of the other player.

In a two-person, zero sum game, a person can win only if the other player loses, no cooperation is possible. Thus, the minimax theorem, which Von Neumann proved in 1928, states that every finite, two-person constant sum-game has a solution in pure or mixed strategies. Specifically, it says that for every such game between player A and B, there is value V and strategies for A and B such that, if A adopts its optimal(maximin)strategy, the outcome will be at least as favourable to A as V, if B adopts its optimal(minimax) strategy, the outcomes will be no more favourable to A than V. Thus, A and B have both the incentive and the ability to enforce an outcome that gives an(expected) pay-OFF of V.

In mixed strategy and minimax theorem, Andrew Colman's game theory and experimental games shows the following example:

In 1943, the allied forces received reports that a Japanese convoy would be heading by sea to reinforce their troops. The convoy could take on of two routes - the northern or the southern rout. The allies had to decide where to disperse their reconnaissance aircraft- in the north or the south- in order to spot the convey as early as possible. The following payoff matrix shows the possible decisions made by the Japanese and the allies, with the outcomes expressed in the number of days of bombing the allies could achieve with each possibility.

By this matrix, if the Japanese chose to take the southern route while the allies decided to focus their racon planes. In the north, the convoy would be bombed for 2 days. The best outcomes for the allies would if they placed their airplanes. In the south and the Japanese took the southern route, the best outcomes for the Japanese would be to take the northern route, provided the allies were looking for them in the south.

To minimize the worst possible outcome, the allies would have to choose the north as the focus of their reconnaissance efforts. This ensures them 2 days of bombing, whereas, they risk only 1 day of bombing if they focus on the south. Therefore, by minimax, the best strategy for them would be to focus on the north.

The Japanese can use the same strategy, the worst possible outcome for them is the 3 days of bombing which might occur if they took the southern route. Therefore, the Japanese would take the northern route.

It is, in fact, what had occurred: both the allies and the Japanese chose the north, and Japanese convoy bombed for 2 days.

The previous matrix had a saddle point, meaning that both the Japanese and allies settled on the (North, North) square as the best outcome for both of them. Neither could do any better if the opponent was rational. In this case, the maximin and the minimax, they would choose the south and surely forfeit one day of bombing.

So, Nash proved that for a certain very broad class of games of any number of players, at least one equilibrium exists – so long as one allows mixed strategies. (Mathematics & Genius & Games) ibid.

So, let`s simplify THE MAXIMIN-MINIMAX PRINCIPLE

THE MAXIMIN- maximizes the smallest gain (minimum payoff)

THE MINIMAX – minimizing the possible loss

THE MAXIMAX- maximizing the maximum payoff (gain)

THE MINIMINI- minimizing the minimum possible loss

Thus, these simple principles could be use like SWOT analysis and Six sigma to improve our strategic efficiency and efficacy. Pareto efficiency optimise the strategy while conditioning exchange efficiency, production efficiency and output efficiency.

So, this principle could have pragmatic application in business CBA (cost benefit analysis) administrative efficiency as well as in good governance via SMART government.

SECTION- SEVEN

Basically, it`s true to hold Darwinism mantra; "it is not the strongest of the species that survives, not the most intelligent, but the one most responsive to change".

"The real test of 'knowledge' is not whether it is true, but whether it empowers us. Scientists usually assume that no theory is 100 per cent correct. Consequently, truth is a poor test for knowledge. The real test is utility. A theory that enables us to do new things constitutes knowledge"

- Yuval Noah Harari

Utility Theory

It Explain, behaviour of individuals based on the premise people can consistently rank order their choices depending upon their preferences. As each individual show different preferences to get highest utility & maximum satisfaction.

– Samarjeet Tripathi

Game Theory does not attempt to state what a player's goal should be; instead, it shows how a player can best achieve his goal, whatever that goal is.

Utilities play a central role in game theory. They capture the preferences that the players have for different outcomes in terms of real numbers thus enabling real-valued functions to be used in game theoretic analysis. So far we have implicitly assumed that utility functions can correctly and faithfully capture the preferences the players have for different outcomes. The utility theory developed by von Neumann and Morgenstern provides a scientific justification for this assumption.

Von Neumann and Morgenstern understood this distinction; to accommodate all players, whatever their goals, they constructed of theory of utility, they began by listing certain axioms that they thought all rational decision makers would follow (For example, if a person likes tea better than coffee, and coffee better than milk, then that person should like tea better than milk). Then they proved that it was possible to define a utility function for such decision makers that would reflect their preferences. In essence, a utility function assigns a number of each player's alternatives to convey their relative attractiveness. Maximizing someone's expected utility automatically determines a player's most preferred option.

Via example, mathematically, expected utility theory: a decision theory for a single agent can be analyses as below -

Planning a party - a game against nature.

Agent is planning a party, and is worried about whether it will rain or not. The utilities and probabilities for each state and action can be represented as follows:

The expected utilities of an action A given uncertainty about a state

S = probability(S/A) *utility(S/A) plus probability (not S/A) utility (not S/A)

Well action A can be viewed as a compound gamble or outcome.

Also, the probability of a state can depends on the agent's choice of action, although, in the above example, it does not.

For the party problem: EU (outside) is equal 2.33;

EU (inside) is equal to value 2

hello dear, did you get, how solution reached. Good then solve it and imagine your own case to apply?

Therefore, choose outside, the action with the higher expected utility.

Problems with the theory of expected utility

(1) Human preferences do not obey the assumptions of the theory

(a) Violations of axioms (transitivity, reducibility, independence)

(b) Violations of invariance (framing effects: reference point dependency and loss aversion, ratio-difference principle)

(2) Assumes there are no rational "opponents" or other intelligent agents who are part of the game.

(Noncooperative) game theory- decision theory for more than one agent, each acting autonomously (no binding agreements)

In the examples, we'll assume two self-utility maximizing agents (or players), each of whom has complete information about the options available to themselves and the other player as well as their own and the other's payoffs (utilities) under each option.

Example - Friends hoping to see each other

Consider two people, Chris and Kim. They both enjoy each other's company, but neither can communicate with the other before deciding whether to stay at home (where they would not see each other) or go to the beach this afternoon (where they *could* see each other). Each prefers going to the beach to being at home, and prefers being with the other person rather than being apart.

Each player has a set of *strategies* [Home , Beach] for both players in this example.

Specifying one strategy i for the row player (Chris) and one strategy j for the column player (Kim) yields an outcome, which is represented as a pair of payoffs (Rij,Cij), where Rij is the utility the row player receives, and Cij is the utility the column player receives.

In this example, going to the beach is a *(strictly) dominant* strategy for each player, because it always yields the best outcome, no matter what the other player does. Thus, if the players are both maximizing their individual expected utilities, each will go to the beach. So Beach-Beach is a *dominant strategy equilibrium* for this game. Because of this, Kim and Chris, if they are rational, do not need to cooperate (make an agreement) ahead of time. Each can just pursue their own interest, and the best outcome will occur for both.

SECTION- EIGHT

Non-Zero-Sum Game or Two Person Variable Sum-Games

"The actual rewards that come from arguing with the other people have nothing to do with winning and losing. A good argument helps us refine our own ideas and discover where our reasoning is the weakest. Other people`s opposition can help us turn our own half-formed ideas into clear assertions backed by solid reasoning. And setting our ideas and opinions against someone else`s help us know each other better, which makes us better friends. We get these benefits from arguments when we collaborate with a partner. We do not get them when we try to destroy an enemy. That is how non-zero-sum games work."

-Michael Austin

Much of the early work in game theory was on two-person constant-sum games, where no player is able to affect the combined payoff. The players in such games have diametrically opposed interests, and there is a consensus about what constitutes a solution (as given by the minimax theorem), most games that arise in practice, however, are variable-sum games or non-constant-sum games; the players have both common and opposed interests.

So, in non-constant-sum games if Player A employs an optimal mixed strategy, Player B can increase his expected payoff by not following the same mixed strategy. The solution lies in either collusion or non-collusion between the two players, the former is known as cooperative non-constant sum game and the latter as non-cooperative non-constant-sum game.

Thus, non-zero-sum game or two-person variable sum-games or non-constant-sum games describes a situation in which the interacting parties‘ aggregate gains and losses can be less than or more than zero. So non-zero-sum games can be either competitive or non-competitive.

In simple, non-zero-sum game can be defined as a situation where one decision maker's gain (or loss) does not necessarily result in other decision maker's loss (or gain). In other words, where the winnings and losses of all players do not add up to zero and everyone can gain: a win-win game.

Problems in the real world do not usually have straight forward result. The branch of game theory that better represent the dynamics of the worlds we live in is called the theory of non-zero-sum games.

Non-zero-sum games differ from zero-sum games in that there is no universally accepted solution that is, there is no single optimal strategy that is preferable to all others, nor is there a predictable outcome. Non-zero-sum games are also non-strictly competitive, as opposed to the completely competitive zero-sum games, because such games generally have both competitive and cooperative elements. Players engaged in a non-zero-sum conflict have some complementary interests and some interests that are completely opposed.

A Typical Example; The Battle of The Sexes is a simple example of a typical non-zero-sum game. In game theory, The Battle of The Sexes (BOS) is two players coordination game that also involves elements of conflict. The game was introduced in 1957 by LUCE and RAIFFA in their classic book, games and decisions.

In this example, imagine that a man and a woman hope to meet this evening, but have a choice between two events to attend, and prize fight and a ballet. The man would prefer to go to prize fight. The women would prefer the ballet, both would prefer to go to the same event rather that different ones.

If they cannot communicate, where should they go?

The wife's payoff matrix is represented by the first element of the ordered pair while the husband's payoff matrix is represented by the second of the ordered pair.

From the matrix above, it can be seen that the situation represents a non-zero-sum, non-strictly competitive conflict. The common interest between the husband and wife is that they would both prefer to be together than to go to the events separately. However, the opposing interests is that wife prefers to go to the ballet while her husband prefers to go to boxing match.

So, the example of Battle of Sexes seems to be an unsolvable dilemma, however, this problem can be solved if either the wife or husband restricts the other's alternatives.

NOTE:

comparatively we must outline related subject matters as below –

1. **ZERO SUM GAME - your loss is my gain: A game in which, winner(s) receive(s) the entire amount of the payoff which is contributed by the loser (strictly competitive). Here, each player`s expected utility in Nash equilibrium must be equal to zero**
2. **TWO PERSON ZERO SUM GAME; a game which involves only two players, say player A and player B, and where the gains made by one equal the loss incurred by the other.**
3. **SIMULTANEOUS MOVES: Lack of information about the opponent`s move**
4. **Pareto optimal outcome:** itis**impossible to make on of the players better off without making another one worse off.**In a non-zero-sum game, a Nash equilibrium is not necessarily pareto optimal.

SECTION- NINE

"21st century global world is no longer a zero-sum game but a multi-dimensional -inclusive arena where cooperation and competition often occurred simultaneously for sustainable climate friendly development."

-Samarjeet Tripathi

Cooperative Versus Non-Cooperative Games

To plan join strategies and cooperate to maximize the pay off toward a new paradigm.

Games which firms play can be either cooperative or non-cooperative, and they are central to many games.

Cooperative games are the ones in which players are convinced to adopt a particular strategy through negotiations and agreement between players, a strategy is complete plan of action a player will take given the set of circumstances that might arise within the game.

However, non-cooperative games refer to the games in which the players decide on their own strategy to maximize their profit.

Communication is pointless in constant-sum games because there is no possibility of mutual gain from cooperation, in variable-sum games, on the other hand, the ability to communicate, the degree pf communication, and even the order in which players communicate can have profound influence on the outcome.

So, if the two firms can sign a binding contract to share the profits between them from the production and sale of carpets, the game is called a cooperative game.

Whereas, in a situation of non-cooperative games while the computing firms take each other's actions into account but they take decisions independently and adopt strategies regarding pricing, advertising, product variation to promote their interests.

It should be noted that a basic difference between a cooperative and non-cooperative game lies in the possibility of negotiating an enforceable contract.

Thus, in cooperative Game; players negotiate binding contracts that allows them to plan joint strategies.

Example: buyer and seller negotiating the price of a good or service or a joint venture by two firms (e.g., Microsoft and apple). So, it is binding contracts possible to reach posterosuperior position for both.

And in non-cooperative game; negotiation and enforcement of a binding contract are not possible.

Example: two competing firms take each other's likely behaviour into account when independently setting pricing and advertising strategy to sign to gain market share.

In globalised era, cooperation among countries can help solve joint problems and share knowledge and best practices.

It often emerges around three common reasons, as per World Economic Forum (WEF) –

- Common challenges that extend beyond national borders;
- Political, cultural, religious and economic commonalities that foster integration
- Inter-country engagement in regional and/or global processes.

SECTION - TEN

The Nash Equilibrium: Cooperative Non-Constant-Sum Game

"Objective facts are Nash equilibrium points in the contest of competing wills"

E.E.E., A Warm Mirror Neuron on A Memory

Game theory, shorn of the mathematics, is probabilistic prediction of how two or parties act in a 'game', that is an interaction between them. It is assumed – and that seems reasonable – that the parties are all rational and wish to maximise the benefit to themselves.

In constant-sum game no player is able to affect the combined payoff, but in non-constant-sum game, if Player A employs an optimal mixed strategy, Player B can increase his expected payoff by not following the same mixed strategy. The solution lies in either collusion or non-collusion between the two players. The former is known as cooperative non-constant-sum game and the latter as non-cooperative non-constant-sum game.

The Nash equilibrium is an example of cooperative non-constant-sum game.

In 1950s, John Nash - The American mathematician later featured in the book and film "A Beautiful Mind" wrote a two-pages paper that transformed the theory of economics. His crucial, yet utterly simple, idea was that any competitive game has a notion of equilibrium:

A collection of strategies, one for each player such that no player can win more by unilaterally switching to a different strategy. For Game Theory, John F. Nash received the Nobel prize for Economics in 1994. It is important to note that all games do not have a dominant strategy, but still the firm achieve equilibrium.

In adoption of their strategy, the application of Nash equilibrium is quite relevant here.

Thus, Nash equilibrium describes a set of strategies where each player believe that it is doing the best it can, given the strategy of other player or players.

In Nash equilibrium, each player adopts his best strategy. A Nash equilibrium is a collection of strategies, one for each player, such that no player would unilaterally alter his strategy. To explain Nash equilibrium, let's take two players who are involved in a simple game of writing words. The games assume that each player writes two words independently on a paper. Player A writes 'top' or 'bottom' and Player B writes 'right and left'. Then the scrutinization of their papers reveals the payoff got by each is, as shown in given Table-1

Make a table of this case study – practice makes perfect

Suppose Player A prefers the top and Player B prefers the left from the Top-Left box of the matrix. It is seen that the payoff to Player A is 2 as the first entry is left box, and payoff the Player B is second entry, 4 in this box. Next if Player A prefers bottom and Player B prefers right then the payoff to Player A is 2 and to Player B is 0 in the bottom-right box.

From the above, we can infer that Player A has two strategies, he can choose either the top or the bottom. From the point of view of Player, A, it is always better for him to prefer the bottom because the choices 4 and 2 are greater than the figures at the top.

Likewise, it is always better for Player B to prefer the left because the choice 4 and 2 are greater than the figures at the right i.e., 2 and 0. Here the equilibrium strategy is for Player A to prefer the bottom and for Player B to prefer the left.

Pay-off matrix for other second case – well, now you are almost perfect and ready to chew more

This matrix reveals that there is one optimal choice of strategy for a player without considering the choice of the other player. Whenever Player A prefers the bottom, he will get a higher pay-OFF of 4 irrespective of what Player B prefers. Similarly, Player B will get a higher pay-OFF of 4 if he prefers left irrespective of what Player A prefers. The preference bottom and left dominate the other two alternatives and hence we get an equilibrium does not occur often. The matrix table-2 shows an example of this particular phenomenon.

In the above matrix when Player B prefers the left, the pay-OFF to Player A are 4 and 0 because he prefers the top. Likewise, when Player B prefers the right; the pay-OFF to Player A are 0 and 2 because he prefers the bottom, when Player B prefers the left 2, Player A would prefer the top 4, and again when Player B prefers the right 4, Player A would prefer the bottom 2.

Here the optimal choice of Player A is based on what he imagines Player B will do. A Nash equilibrium can be interpreted as a pair of exceptions about each player's choice such that when the other player's choice is revealed in the above matrix, the strategy top-left is a Nash equilibrium. In a Nash equilibrium, no player has an incentive to depart from it by changing his own behaviour.

Thus, in the cooperative non-constant-sum game, the most rational thing for the two player is to collude and thus to increase their combined pay-OFF without reducing any one's pay-off. So, Nash equilibrium tries to arrive at a "Fair Division" by evaluating the pay-OFF for both players.

So, In Nash equilibrium theory, a player can achieve the desired outcome but not deviating from their initial strategy. every player wins because everyone gets the outcome they desire.

Existence of a Nash Equilibrium

Not every strategic game has a Nash equilibrium, as the game Matching pennies shows the case.

Existence of Nash equilibrium is a key question investigated extensively in game theory. For two person zero-sum games with finite strategy sets, we have seen in the previous chapter, the *minimax theorem*, which establishes the existence of at least one mixed strategy equilibrium. John Nash, in his brilliant work, generalized the notion of an equilibrium to games with three or more players and also established the existence of at least one mixed strategy Nash equilibrium for every finite strategic form game. Nash equilibria in games turn out to be fixed points of appropriately defined mappings. In fact, the existence of equilibria in games is closely coupled with fixed point theorems such as *Brouwer's fixed point theorem* and *Kakutani's fixed point theorem*

Thus, a non-cooperative game is said to be in Nash equilibrium if no player has incentive to change his individual game strategy after considering the strategies of all other players. The prisoner`s dilemma is a classical example of the Nash equilibrium. As a reminder, Prisoner`s dilemma is a situation in which two prisoners are convicted as accomplices in a crime. The prisoners are placed in solitary confinement, so they have no method of communicating with each other. They are then each presented with the following proposal:

i. If they both confess, they will each spend 8 years in jail.
ii. If only one of them confess, he will be set free while the other will spend 10 years in jail.
iii. If neither of them confess, they will each spend 1 years in jail.

This game is in Nash equilibrium when both prisoners confess. Why? Because under these circumstances, neither prisoner benefits by changing his strategy. if prisoner 1 were to change his strategy and instead kept quiet, then he would receive a longer jail sentence than he would if confessed. And prisoner 2 will be able to walk away without punishment. Prisoner 2 should maintain his strategy as well by the same logic. Although the best strategy for the group as a whole would be for both to keep quit, individually the prisoners are better off confessing since they have no way of knowing the other prisoner`s strategy beforehand, and staying quiet while the other confesses would result in 10 years of jail time.

The Nash equilibrium can be applied to a variety of real-life situations. It explains, for example, why we overfish the sea: although overfishing is clearly bad for the ecosystem as a whole, it would be bad for an individual company to stop fishing because then that company would stop profiting while other companies continue to fish and, hence, continue to make profit. Same case with coal mining, sand mining and other natural resources over-exploitation cases. The Nash equilibrium can also be applied in economics, war, politics, environment, everyday life and countless other fields.

Nash equilibrium application in administrative, political and judicial system -

As we already discussed important concept related to Nash equilibrium with many prospective applications in many areas. Well, in administrative efficiency Nash equilibrium could have pragmatic effect. In my upcoming book on GAEORY, I will discuss in length about the relationship between Nash equilibrium and political, administrative and judicial efficiency.

Take a deep breath, so you could have few exercises.

Now Brainstorming time:

- What is the role of Nash equilibrium in our **everyday life**.?
- How could Nash equilibrium accelerate the efficiency of bureaucracy?
- How our judiciary system could absolve from prisoner`s dilemma so it could ease efficiently from piles of pendency cases.

SECTION- ELEVEN

The Prisoner's Dilemma: Non- Cooperative Non-Constant-Sum Game

A situation where two parties, separated and unable to communicate , must each choose between cooperating with the other or not.

"Marriage is a prisoner`s dilemma in which you get to choose the person with whom you`re in partnership. This might seem like a small change, but it potentially has a big effect on the structure of the game you`re playing. If you knew that, for some reason, your partner in crime would be miserable if you weren`t around- the kind of misery even a million rupees couldn`t cure – then you`d worry much less about them defecting and leaving you rot in jail."

-Samarjeet tripathi

If collusion is ruled out, we enter the real me of non-cooperative non-constant-sum game where each player acts on his guesses about other's choice of strategy.

Non-Cooperative-sum games(non-constant) may be variety of types. The two players guided by self-interest, as they are likely to be, may select strategies which may be mutually harmful.

The firm working in oligopolistic market make decisions in the face of uncertainty about how their rivals will react to their moves. As we know, game theory is mathematical technique of analysing the behaviour of rival firms with regard to changes in prices, output and advertisement expenditure in the situations of conflicts of interest among individuals or firms. An important game model that has significant implications for the behaviour of the oligopolists is popularly known as prisoner's dilemma, model of prisoner's dilemma explains how rival behaving selfishly act contrary to their mutual or common interests.

Prof. Trucker's "prisoners dilemma" is an interesting case of a non-cooperative non-constant-sum game. Suppose two persons A and B have been arrested for a joint crime. Each is brought for interrogation separately and no communication is allowed between them.

They are given the following alternatives:

- If they do not confess, each will get 1year sentence
- If either A or B confess, he goes free but the partner gets 7-years sentence
- If both confess, each gets 3-years sentence

What A and B should do, imagine a case and make a matrix of given below analysed case study?

First, let us consider A's dilemma should he confess or not confess irrespective of what B does? If he confesses and B also confesses, he gets 3-year sentence. If B does not confess, and A confess, B gets 7-year sentence and A is set free. On the other hand, B also in the same dilemma, if he confesses and A does not confess, A gets 7-year sentences and B is set free. If both do not confess, each gets 1-year sentence.

In order to get the least sentence, each follows the followings strategy:

A think that if he confesses, either he gets 3-year sentence or goes free. If he does not confess, he gets either 7-year or 1-year sentence. So, the best strategy for him is to confess.

B also thinks on the same lines. By confessing, he either gets 3-year sentence or goes free.

By not confessing, he gets either 7-year or 1-year sentence. So, he also chooses to confess. Thus, both A and B confess and each gets 3-year sentence.

This is the dominant strategy in the prisoner's dilemma whereby each prisoner expects the worst from the other by confessing and the period of their sentences s equal, that is 3-yeas. This mutual confession strategy denotes a Nash equilibrium, but both would prefer by keeping silent (or not to confess) and get 1year sentence.

How can we avoid Prisoner`s dilemma

in this classic game theory experiment, you must decide: rat out another for personal benefit, or cooperate? The answer may be more complicated than we think.

Everybody likes to operate in an environment of trust. When you deal with people you trust, things get done faster, stress is reduced and new opportunities open up. As E.M. Forster once wrote, "One must be fond of people and trust them if one is not to make a mess of life."

Many businesses are able to avoid prisoner`s dilemma. for example, Amul Dairy`s has maintained trustful partnerships with its suppliers for decades, TATA India Inc herself synonymous to TATA TRUST.

Traditionally, the most effective option for overcoming a prisoner`s dilemma is **the tit for tat strategy**, in which you start out cooperating and then replicate whatever the other player`s last move was. so if he cooperates, you do the same, if not, you retaliate. It`s simple and long line of experiments rule. Well, this solution also become less practical in real world.

A more viable strategy is to **network their organization**by creating personal relationships from disparate groups through embedding, combined training, liaison programs and promotion policies. research into a network.

Further, **research into a network**finds that it takes relatively few connections to drastically reduce social distance, so a networking strategy viable even for large organizations. Even if two people have never met, a mutual relationship with a trusted third party can help close the gap quickly.

SECTION- TWELVE

"The power to constrain an adversary depends upon the power to bind oneself."

- Thomas Schelling

Theory of Moves

In theory of moves Steven Brams attempts to provide an alternative to standard game theory that more easily captures the dynamic aspects of strategic behavior.

Adding a dynamic dimension to game theory allows players to look ahead before making moves, thereby creating a more realistic game.

Another approach to including cooperation in prisoner's dilemma and other variable-sum games is the theory of moves (TOM) proposed by the American political scientist Steven J. Brams. TOM allows players, starting at any outcomes in pay-OFF matrix, to move and countermove within the matrix, thereby capturing the changing strategic nature of games as the evolve over time. In particular, TOM assumes that players think ahead about the consequences of all of the participant moves and countermoves when formulating plans. Thus, the theory of moves combines the extensive form and normal forms of classical game theory, a theory of moves game is played on a pay-OFF matrix, like a normal form game. The player, however, can move from one outcome in a pay-OFF matrix to another, so the sequential moves of an extensive-form game are built into the more economical normal form.

Well, it is dynamic because players do not make choices de novo.

Instead, their choices depend on the past and present as well as future, which players can anticipate at least in part and they can make rational calculations.

The theory of moves includes six basic rules:

Rule1-states that a game starts at an initial state, which is a row-and-column intersection of a pay-OFF matrix.

Rule- 2 says that either player can switch to a new strategy, thereby generating a new outcome, the first player to more is called Player 1.

According to **Rule 3**, the other player, Player 2, can then move.

A game's end is determined by **Rule 4**, the player respond alternately until neither switch's strategies, the resulting outcome is the "Final State", which is the only point at which the players accrue pay-OFFs.

The remaining rules, can be call rationality rules, explain the reasons for moving or not moving.

Rule 5states that a player will not make a move unless it leads to a preferred outcome, based on his or her anticipation of final state.

Rule 6– that is, two sidedness rules, says that a player considers the rational calculations of other player before moving, taking into account their possible moves, the possible countermoves of the other players, their own counter-countermoves, and so on. Thus, a player may do immediately better by moving first according to Rule 5; but if this player can do even better by letting the other player move first, and it is rational for that player to do so, then first player will await this move, according to Rule 6.

Some of the differences between classical game theory and the theory of moves arise in an imaginary confrontation situation called a **Truel.**

The theory of moves offers a different perspective. Instead of assuming simultaneous strategy choices, it asks each player: given your present situation and situation that you anticipate will ensure if you fire first, should you fire? At the start of a Truel, all the players are alive, which satisfy their primary goal of survival but not their secondary goal of surviving with as few others as possible.

To sum up, the theory of moves (TOM) renders game theory a more dynamic theory, by postulating that players think ahead not just to the immediate consequences of making moves, but also the consequences of the counter move to those moves, but also counter-countermoves, and so on, it extends the strategic analysis of conflicts into the more distant future.

TOM has also been used to elucidate the role those different kinds of power- moving, order and threat-may have on conflict outcomes, and to show how misinformation can affect player choices. These concepts and the analysis have been illustrated by numerous cases.

For brief analysis,let's take an example of Cuban missile crisis original case study taken by PROF. Steven J. Brams.

Hello dear reader! Do a brainstorm and make a matrix of this case study?

As we know, **the Cuban missile crisis, also known as October crisis of 1962,**The Caribbean crisis or missile scare, was a 1 month, 4-day confrontation between the US and Soviet Union which escalated into an international crisis.

In compromise, they get nonmyopic equilibrium; in pay-OFF 4 is best, 3 next best and so on as 1 is worst.

The cell of this pay-OFF matrix of the Cuban crisis represents possible states of the world (or outcomes). The ordered pair in each cell of the matrix reflects Bram's understanding of the relative ranking of the four possible outcomes by united states and Soviet Union, respectively. Brams assumes that once a game begins either player can move from whatever outcomes is the initial state (or states 940 outcome) and, if it does, the other can respond, the first can counter- respond, and so on. The process continued until one player decides not to respond, and the outcome that they are at is the final outcome.

As in this case, when Kennedy (US President) announced the blockade on October 22, the Soviet missiles were already being installed in Cuba, this, the initial states of the world, was the worst for the United States and best for the Soviet Union. Thus, this outcome is labelled "Soviet Victory". But at end removal of the soviet union's nuclear missile from Cuba and removal of USA nuclear missiles from Turkey and Italy.

Brams, suggests several reasons why the Soviet Union would then withdraw the missiles and induce its next best outcomes, the compromise outcome, rather than stick with its initial choice. So, because, moving power ability to take finally in pay-OFF either at (3,3), or force it to choose between (3,3) and (4,1).

SECTION- THIRTEEN

N - Person Games

...in a world with many multiple interests, sometimes in conflict and sometimes in cooperation with one another, all the facts and quantifications by themselves do not necessarily point unambiguously to a "correct" and "fair" is a matter of judgement in a multipolar world. That this is the case is made particularly clear by the theory of n-person games.

-L. S. Shapley

In game theory, an n-player game is a game which is well defined for any number of players. This is usually used in contrast to standard 2-player games that are only specified for two players. Thus n-person games allow for any (Finite) number of players.

One example of n-player prisoner's dilemma is the Diner Dilemma. Well, the unscrupulous diner's dilemma, the situation imagined is that several people go out to eat, and before ordering, they agree to split the cost equally between them. Each diner must now choose whether to order the costly or cheap dish. It is presupposed that the costlier dish is better than cheaper, but not by enough to warrant paying the difference when eating alone. Each diner reasons that, by ordering the costlier dish, the extra cost to their own bill will be small and thus the better dinner is worth the money. However, all diners having reasoned thus, they each end up paying for the costlier dish, which by assumption is worse than had they each order the cheaper.

Many methods there to find out equilibrium in n-person game. The branches of mathematics on which game theory especially N-person games leans heavily is called theory of sets.

So, one may define a concept of an n-person game in which each player has a finite set of pure strategies and in which a definite set of payment (Factor) to the n players corresponds to each n-tuple of pure strategies, one strategy being taken for each player.

So, N-person game theory examines the relationship between the characteristic function of a game and various combinations of payoffs and coalitions.

Let a three-person game yield no advantage to solitary action but reward cooperation between any two players or among the three players with a joint gain of one unit.

N - person theory, then, is concerned with things like cooperation, coalition, organizational structure, commitment, trust, compromise, threat, enforceability, and indeed the whole legal/social/cultural environment. It deemphasizes questions of tactical optimization, the detail spelling out of rules, and the numerical calculation of outcome and payoffs. nevertheless the concept is heavily mathematically.

SECTION- FOURTEEN

Sequential and Simultaneous

In game theory, games are divided into group, two of which are simultaneous and sequential games. In sequential games, each player knows the movements of other players, the players continue the game one after another. In simultaneous games, players play simultaneously and are unaware of the movements and actions of other players. Chess is a sequential game and stones, paper, scissors is a simultaneous game. Even, many of today's popular games are in the group of simultaneous games. Call of the Duty is one of these games which has many fans.

Thus, sequential or dynamic play is a game in which the player chooses has action before others choose their action. In this type of game, each player is informed of the move made by the opposing player and according to that, he decides what to do in the next move. In fact, in such games the next move of each player depends on the previous move of other players. So, it's important which player starts the game and this initial move. In the game can be a source of player gain in the game in the game and the opponent's loss. Sequential game is predicate unlike simultaneous. Chess is an example of sequential game.

Whereas in simultaneous or static play, players continue to play simultaneously without knowing each other's movements and actions. In this type of game, players do not have much information about the performance of their opponent and chooses the game strategy regardless of opponent. This type of game is the opposite of a sequential game in which the player continues the game according to the actions of the opponent.

So, via examples the concept of simultaneous games can be discuss as the following –

Stone, Paper, Scissors: In this game, all players play at the same time and the choice of the rest has no effect on their choice. For example, if the first player chooses the paper, the next player will no doubt choose the scissors.

Voting in Elections: In elections, each person makes his choice regardless of the votes of the others.

Thus, in simultaneous games, because each player plays independently of the strategy and movements of the other players, it does not matter which player starts the game first, and it has no effect on the players win or lose.

As in simultaneous play, the player has no knowledge of the other players decisions, so he takes a lot of risks. Thus, this game is unpredictable unlike consecutive games.

Prisoner's problem game can be taken as part of simultaneous game because neither prisoner is aware of the other's decision.

Another example for simultaneous game is the game of – The Battle of Sexes (we already discussed in non-zero-sum game). Even the game of wire's war, in which a couple wants to go to the theatre or cinema, and in this game, it is important for them to be together and they have to decide at the same time. Well, other most important issues in simultaneous game is the Nash equilibrium issue.

Well, in contrast to the linear chain of reasoning for sequential games, a game with simultaneous moves involves a logical circle. Although players act at the same time, in ignorance of other players` current actions, each is forced to think about the fact that there are other players who in turn are similarly aware ... his best action is an integral part of the calculation. (Games & Poker)

SECTION- FIFTEEN

Power in Voting: The Paradox of The Chair's Position

Many applications of n-person game theory are concerned with voting in which strategic calculations are often rampant.

'Chairman's Paradox' that was first noticed by Farquharson in this path breaking tracts on sophisticated voting, Theory of Voting (1969).

In sort, chairman's paradox is concerned with the case of a three-member committee in which a particular player who has a regular and tie-breaking vote – the chairman's- not only will do worse in specific instances under the plurality procedure for three alternatives than if he did not have such a vote but will also worse overall. That is, the chairman's a priori probability of success (getting what one wants) for all possible games with linear (strict) preference order is lower than that of the regular members. It is demonstrated that this, result which comes about if voters act strategically rather than sincerely, is not as robust as it has been thought to be. By merely replacing the standard assumption of linear preference orders with weak preference orders, the priori success of the chairman is now greater than that of the other two players. We also point to a new paradox of sophisticated voting.

SECTION- SIXTEEN

The Von Neumann and Morgenstern Theory

Von Neumann and Morgenstern were the first to construct a cooperative theory of n-person games.

According to their game theory, an oligopolist while choosing his strategy will assume that his rivals will adopt a strategy which will be worst for him, that is, rival will adopt the policy which will be most unfavourable to him. That is to say, an oligopolist will adopt the policy of "playing it safe". Given the assumption, from among those strategies which provide him with various minimum gains, an oligopolist will select that one which is maximum in those minimum gains.

VNM defined the solution to an n-person game as a set of imputations satisfying two conditions:

1. No imputation not in the solution dominates another imputation in the solution and
2. Any imputation not in the solution is dominated by another one in the solution.

A VNM solution is not a single outcome but, rather, a set of outcomes, any one which may occur. It is stable because, for the members of the coalition any imputation outside the solution is dominated by- and is therefore less attractive than an imputation within the solution. The imputation within the solution is variable because they are not dominated by any other imputations in the solution.

The significance of Von Neumann`s minimax theorem?

Given zero-sum games between two players it actually provided a way "to solve" them to see what strategies would be adopted by them i.e., how should they play the game. Minimax theorem is basically a principal/abstraction/mathematical model stating how the player behave and is solved by solving a linear program that describes it. Oskar Morgenstern realized that this is a brilliant way to look at economic decision making and convinced Von Neumann to write a book: "Theory of games and economic behaviour" – the book very foundation of game theory.

For zero sum games Minimax equal to Nash equilibrium equal to Stackelberg equilibrium. at that time it was a quite profound insight. Von Neumann created this game based view of looking at the world, described computing the strategies by solving a corresponding linear program, gave birth to the theory of duality (of LP) as a side effect and described how players "tend to act" in one master stroke.

SECTION- SEVENTEEN

The Banzhaf Value in Voting Game

In chairman's paradox model, it was shown that power defined as control over outcomes in not synonymous with control over resources, such as a chair's tie-breaking vote. The strategic situation facing voters intervenes and may cause them to reassess their strategies in light of the additional resources that the chair possess. In doing so, they may be led to "gang up" against the chair, not power to relish.

American Attorney John F. Banzhaf III proposed that all combination in which any player is the critical voter- that is, in which a measure passes only with this voter's support- be considered equally likely. The Banzhaf value for each player is then number of combinations in which this voter is critically divided by the total number of combinations in which each voter (including this one) is critical.

This view is not compatible with defining the voting power of a player to be proportional to the number of votes he casts, because votes per se may have little or no bearing on the choice of outcomes. For example, in a three-member voting body in which A has 4 votes, B has 2 votes, and C has 1 vote, members B and C will be powerless, if a simple majority wins. The fact that member B and C together control 3/7 of the votes is irrelevant in the selection of outcomes, so these members are called dummies. Member A, by contract, is a dictator by virtue of having enough votes along to determine the outcome. A voting body can have only one dictator, whose existence renders all other members dummies, but there may be dummies and no dictator.

A minimal winning coalition (MWC) is one in which the substruction of at least one of its members renders it losing. To illustrate the calculation of Banzhaf values, consider a voting body with two 2-vote members (distinguished as 2a and 2b) and one 3 vote member in which a simple majority wins, there are three distinct MWCs – (3,2a), (3,2b), and (2a,2b) - or combinations in which some voter is critical, the grand coalition, comprising all three members, (3,2a,2b), is not an MWC because no- single member's defection would cause it to lose.

As each member defection is critical in two MWCs, each member's proportion of voting power is two-sixths, or one-third.

Thus, the Banzhaf index, which gives the Banzhaf values for each member in vector form, is (1/3, 1/3, 1/3). Clearly, the voting power of the 3-vote member is the same as that of each of the two 2-vote members, although the 3- vote member has 50 percent greater weight (more votes) than each of the 2- votes members.

The Banzhaf index can be understood via EEC example. Well, in 1958 six west European countries formed the European Economic council (EEC). The three large countries (west Germany, France, and Italy) each had 4 votes on its council of ministers, the two medium -size countries (Belgium and Netherlands) 2 votes each, and the one small country (Luxembourg)1 vote. The decision rule of the council was a qualified majority of 12 out of 17 votes, giving the large countries Banzhaf values 5/21 each, the medium – size countries 1/7 each, and amazingly – Luxembourg no voting power at all.

From 1958 to 1973- when the EEC admitted three additional members – Luxembourg was a dummy. Luxembourg might as well not have gone to council meetings except to participate in the debate, because it's one vote could never change the outcome.

UNIT - II

DILEMMA CAPTIVUS EST SCRIPTOR: A CONCEPTU UNIVERSALI

"The road to hell is paved with intractable recursions, bad equilibria, and information cascades"

: Brian Christian

CHAPTER THREE

CHAPTER 3

The only winning move is not to play: A dilemma

Prisoner's dilemma: a universal concept–

The basic premise of the prisoner's dilemma is that two suspects are placed in two different rooms, and each is asked separately whether or not his partner is guilty. Prison sentences depend on how each suspect responds: if both remain true to each other, they each serve only six months. If both betray each other, they both serve two years. If one betrays and one stays silent, the silent/cooperative partner serves 10 years while his betrayer goes scot-free. As one of the suspects, if you choose to betray, you will have better results, either serving two years or going free. If you choose to cooperate, you're in the cooler for either six months or 10 years. With those odds, most people should betray, even though the best result for everyone would be full cooperation.

"In laymen`s language, the prisoner`s dilemma is a case where two people have to choose to cooperate with each other or betray the other. The game of prisoner`s dilemma is of important relevance to the market structure especially oligopoly, as prisoner`s is a game that explains why it may be hard to maintain cooperation including when there are mutual benefits. For example, in case of pandemic (Covid-19), the cooperation of government and public at a large."

In this pandemic situation, both cooperation is very important to tackle with the situation like government rules regarding Covid-19 appropriate behaviour if follow by public sincerely, the benefit will be at a large. Thus Prisoner`s dilemma is an interesting case of a non-cooperative non-constant-sum game as per PROF. Tucker.

Another real-life situation involving the benefits of cooperation is an arms race between two countries that spend billions of dollars on making weapons for fear that their enemy may be investing in the same defensive strategy. Both lose, considering that the money could have been spent on more productive means. According to Perc's results, countries who can take the risk of forming and trusting allies will aid one another through cooperation. They can more readily cooperate, according to the double resonance phenomenon, by beginning to cooperate during instances of decreased risk, as well as by taking advantage of currently existing long-distance bonds that provide a mechanism for cooperative alliances across the globe.

In Joseph Heller`s novel Catch-22, allied victory in World War ll is a foregone conclusion, and Yossarian does not want to be among the last ones to die. His commanding officer points out, "But suppose everyone on our side felt that way? "Yossarian replies,"then I`d certainly be a dammed fool to feel any other way, wouldn`t I?" Yossarian`s dilemma is just a multi-person version of this. His death is not going to make any significant difference to the prospect of victory, and he is personally better off alive than dead.so avoiding death is his dominant strategy.

How might such dilemma be resolved?

Well if the relationship of the players is repeated over long time horizon, then the prospect of future cooperation may keep them from finking that is tit-for-tat strategy, mixing move, commitment (Thomas Schelling pioneered this concept), Information and Incentives(an in information age, Information play crucial roles; Nobel Laureates ***Goerge Akerlof, Michael Spence, and Joseph Stiglitze*** testifies to its importance) and Aligning interest Interests, Avoiding Enrons could be some ways to resolved dilemmas.

Game theory and Information economics (the economics of Information) have given us valuable insights into these issues.

SECTION- ONE

The Relevance of Prisoner's Dilemma in Oligopoly Market

Firms in an oligopoly can increase their profits through collusion, but collusive arrangement are inherently unstable.

Oligopoly is a market structure in which there are few firms producing a product. when there are few firms in the market, they may collude to set a price or output level for the market in order to maximize profits. As a result, price will be higher than the market - clearing price, and outputs is likely to be lower. At the extreme, the colluding firms may act as a monopoly, reducing their individual output ton that their collective output would equal that of a monopolise, allowing them to earn their profits.

E.g., OPEC: the oil producing country of OPEC have at times, cooperated to raise world oil prices in order to secure a steady income for themselves

The firms working in oligopolistic markets make decision in face of uncertainty about how their rivals will react to their moves.

Game theory provides a framework for understanding how firms behave in an oligopoly. An important game that has significant implications for the behaviour of oligopolist is popularly known as prisoner's dilemma. Model of Poisoner's dilemma explains how rivals behaving selfishly act contrary to their mutual common interest. Thus, the prisoner's dilemma is a type of game that illustrates why cooperation is difficult to maintain for oligopolists even when it is mutually beneficial in this game, the dominant strategy of each actor to defect.

However, acting in Self- interest leads to a sub- optimal collective outcome, well, game theory is generally not needed to understand competitive or monopolized markets.

IN an oligopoly, Firms are interdependent; they are affected not only by their own decision of others firms in the market as well.

Game theory offers a useful framework for thinking about how firms may act in the context of this interdependence, more specifically game theory can be used to model situations in which each actor, when deciding on a course of action, muse also consider how others how others might respond to that action.

For example, game theory can explain why oligopolies have trouble to maintain collusive arrangements to generate monopoly profits. While firms would be better off collective they cooperate, each individual firm has a strong incentive to cheat & undercut their competitors in order to increase their market share. Because the incentive to defect is strong, Firms may not even enter into a collusive agreement if they don't perceive there to be a way to effectively punish defectors.

The prisoner's dilemma basically provides a framework for understanding how to strike a balance between cooperation and competition and is a useful tool for strategic decision making.

As a result, its kinds application in diverse areas from business, finance, economics and political science to philosophy, psychology, biology and sociology.

If collusion is ruled out, we enter the realm of non - cooperative non-constant sum game where each player acts on his guess about other's choice of strategy.

Non- cooperative non -constant - sum games may be a variety of types., the two players guided by self-interest, as they are likely to be, may select which may be mutually harmful.

Prisoner's dilemma is west example of non - cooperate non- constant sum games.

The game of prisoner's dilemma is of important relevance to the oligopoly theory. The incentive to cheat by a number of a Cartel (e.g., in modes of collusive oligopoly, and eventual collapse of cartel agreement is better explained with the model of prisoner's dilemma instead of two prisoners we take the two firms A and B which have entered into cartel agreement and have fixed the price of the product each has to charge and output each has to produce and sell (i.e. share of the market) the choice problem facing each member firm of the cartel is whether to cooperate and abide by the agreement and thus sharing the make higher individual profits . but if both cheats and violate the agreement, the cartel would break down and violate the agreement, the cartel would break down and profits, would fall to the competitive level.

Imagine payoff matrix for cartel members –

A cartel is an oligopoly in which the members try to collude to behave as a monopoly by setting prices and output to maximize the collective profit. The outcome for a cartel is a prisoner's dilemma with a Nash equilibrium with each member doing the best it can, given the behaviour of the others.

It will be scan from the imagined payoff matrix, that if both firms cooperate and abide by cartel agreement, they share monopoly profits; ?15 lakhs to each of them (right and bottom).

If both firms cheat and thus violate the agreement, profits to each firm fall to the competitive level ?5 lakh to each firm (left hand top)

If both firms cheats, while firm B cooperates , profits drop to low level of ?2 lakh and a's profits , rise to ?25 lakhs (left and bottom) , on the other hand , if firm B cheats and firm A adheres to the agreement , profits of a declines to ?2 lakh and b's profits shoot up to ?25 lakhs (left hand top).

Well, I hope, you practiced well while in imagination – oh, so which game is this? Let me know

Take an other example, to understand it well:

Venezuela and Nigeria have created an oil cartel. If they cooperate, the potential profit is $10 million dollars, evenly divided. If they both cheat, the maximum profit is only $8 million. Although cooperation is the best outcome, there is a strong incentive to cheat to take advantage of the monopoly price. Examine the matrix on the left from Venezuela's perspective: Venezuela's optimal strategy is to cheat, regardless of what Nigeria does. If Nigeria cheats, Venezuela should cheat to get $4 million in profit; if Nigeria cooperates, Venezuela could get $6 million in profits.

So, Nigeria faces the same choices as Venezuela so its best strategy is also to cheat. The best outcome is in the lower right hand box of the matrix. This, however, is not a stable outcome. The outcome is indeterminate except that it will be a Nash equilibrium.

SECTION- TWO

Strategy models of prisoner`s dilemma -

In the prisoner`s dilemma game we are studying, the strategy of other players is unknown while the profit function and choice space are clear. Our strategy is in this game is to reduce the unclear information and maximize the profit in expectation.

There are atleast three challenges as follows:

I. The strategies of other players are variable and unstable. They may make a different choice in the same situation that`s no best choice can be selected in a single game.
II. All of the other players in the game will have different strategy. and they play together in one game.
III. The performance and efficiency of the strategy should be promised, especially in the case that the number of players is large.

A prisoner`s dilemma is a decision - making and game theory paradox illustrating that two rational individuals making decisions in their own self intrest cannot result in an optional solution. The paradox was developed by mathematicians M. flood and M. Dresher in 1950, and the modern incorporation was conceptualized by Canadian mathematics A. W. Tucker.

The prisoner's dilemma may be expressed as an approach where individual parties seek their welfare at the expenses of the other party, generally, since both participations avoid cooperation in the decision -

making process, they end up in a much worse condition. In the prisoner's dilemma theory, it is the responsibility of the two parties to choose whether to collaborate or not, either party is given to chance to defect, despite the option of the other party. the outcome of the prisoner`s dilemma is either beneficial or injurious to society. Making better economic choices require cooperation between individuals.

So, prisoner's dilemma game can be used as a mode for many Real world situation involving cooperative behaviour. **Let's briefly discuss original prisoner`s dilemma case study:**

As we acclimatized that, the prisoner's dilemma is a problem in the game theory in which two competing players end up in a worse situation because they assume the other one won't cooperate.

Scenario

The police have captured two crimes suspect and are interrogating them in separate rooms, so they can't communicate with each other. The police officers offer both suspect the opportunity to either remain silent or blame another suspect, if both suspects remain silent, they both will serve only three years in prison. If one of the suspect blames another and the other remains silent, the suspect who remained silent would serve five years in prison, while another suspect would be set free.

They offer each the following deal:

- If suspect A snitches on suspect B, A goes free and B spends three years in jail.
- Similarly, if B snitches on A, B goes free and A spends three years in jail.
- If neither of them snitch on each other, then they both spend two years in jail.

The prisoner's dilemma is a great way to understand and analyse human behaviour. In the mid1900s, it would have been relatively easy to predict what each prisoner's choice would be , able it not always right, that's because economists at that time believed in the idea of a homo economicus, or a human that made rational decision purely based on self-interest, all because the individual payoff are higher for each one. However, it's interesting to observe that the prisoner's pursuing of logical, individual reward yields worse results than if they both cooperated and stayed silent. In reality, humans deviate from the expected path of the "economics man" - they display a bias towards more cooperative behaviour in these kind of game, much more than predicted by simple models of self - interested action.

In short, in this game, two individuals determine cooperation or defection. If the two individuals are mutually cooperative, they both earn incomes R; if they defect to each other, the incomes of both sides are P; if one individual is cooperative while the other is in a state of betrayal, the cooperative one gains T. here, descending order of gain as, T – R – P – S and 2R is greater than mutually S and T gain. That mean, total income of the two cooperative individuals is always larger than that gained in case of one individual`s treachery. However, with regard to individuals, the incomes earned by defection to cooperation is greater than that by cooperation to cooperation.

Several typical strategy models are worth to discuss in use to prisoner`s dilemma –

- **TFT (tit-for-tat) strategy,** that is, "return like-for-like" strategy. the main idea of this strategy is that, by starting with cooperation, the strategy selection of a round is made on the basis of the selection from the previous round.
- **PTFT, an improved TFT strategy,** an improved TFT strategy, this strategy is relatively more selfish than TFT. it still starts with cooperation; however, in the following rounds, cooperation is only selectable in the case of an absence of defection for three rounds.
- **GTFT, another improved TFT strategy:** its strategy allows a certain probability of cooperation in the case of rival defection and a certain probability of defection and a certain probability of defection in the case of cooperation. It solves the deadlock arising from mutual defection in the competition.
- **Pavlov,** a different strategic concept. It bullies the weak and fears the strong; namely, cooperation is continued in cases of mutual cooperation. However, defection is selected when one side chooses defection, cooperation is given priority. Such a strategy represents a local optimum in genetic algorithm terms.
- **Random,** random strategy, that is, randomly returning to cooperation or non-cooperation: in related programs, a 50:50 random strategy is more commonly applied; that is, the probability of returning to cooperation or defection is 50%. This strategy is mainly adopted to assess fixed strategies, set competition parameters, and so forth.

- **Normal,** a strategy mode developed by simulating common players: in this strategy, cooperation or defection would be selected with different probabilities based on the selections of both sides in the previous game. This strategy simulates player participation in a game using different strategies.

So, we've gone over what the prisoner's dilemma and brief relevance in oligopoly, while the real life application might seem a little contrived, the payoff matrix is very real and very applicable.

CHAPTER FOUR

CHAPTER 4

Real Life Examples

"I use the game theory to help myself understand conflict situations and opportunities"

-Thomas Schelling

In laymen's language, the prisoner's dilemma is a case where two people have to choose to cooperate with each other or betray the other, each case has different rewards and punishment, if one other gets a more serve punishment. If both people cooperate, then they got small punishment. If both betrays each other they got more punishment. If one betrays and the other cooperates, he gets to go free and the other gets the worst punishment.

You are always punished less for choosing to betray the other person. However, as a group, both of you fare better by cooperating. Let`s simplify the prisoner's dilemma case in yearly scenario.

Well, when we play a second time, the situation becomes interesting.

If you get a reputation for cooperating, people will more likely to cooperate with you. But you risk a worse punishment if you get a reputation for betraying, the more people who may betray you, the more likely you will choose betrayal. The relevancy of prisoner's dilemma is very wide, not only in oligopoly market but also other markets as well. It has very wide application in real world also, for example:

A pair working on a project. You do best if your competitors does all the work, since you get the same grade. But if neither ok you do the word, you both fail. In advertising, if both companies spend more money on advertising, their market share won't change from if neither does. But if one company outspends the other, they will receive 9 benefits.

A sociologist example would be that it's always better to not clean a shared room, since the other might clean it. Depends on the emotional pay-off for cleaning the room for the other, which also depends on the history of the other one cleaning the room every now and then.

Other example would be when overfishing is bad for every Fisherman involved, but for only fisherman it's always better to fish more, irrespective of the choice of the other fisherman.

Once there's a law that punishes overfishing, the pay-offs will change, and things might change in positive way.

Well, take an example of driving, imagine you're driving down a road. The traffic eight ahead is broken or some technical issue.

You are doing 50 mph (~80 Kph for Indians).

On the road orthogonal to yours, you see another car, speeding towards the same intersection.

If you both maintain your speed, one of you will probably t- bone the other.

What do you do, that's dilemma is a prisoner's dilemma. You can defect and keep driving or cooperate and slowdown. This sort of dilemma would happen all the time without traffic lights, even when we are not taking traffic rules seriously. So those traffic signals and strict traffic rules force us to cooperate in turns.

Preventing the unwanted Nash equilibrium of t- boning other cars. Even in cases of broken traffic signals most people default to the good Nash equilibrium of slowing down.

Thus, this theory applies in a oligopoly market where, there is a cut throat competition between two big players in the industry. Each firm adopts the best strategy keeping in consideration the reaction of the rival firm, a price reduction is immediately followed by the rival firm. Continuous price decrease leaves the industry demand curve elastic.

A price increase is not followed by the revival. This leaves the industry demand curve in classific beyond a certain point.

Before, discussing the real – world application of Prisoner's dilemma; Let's have brief idea to 'the oligopoly version of the prisoner's dilemma'.

The members of an oligopoly can face a prisoner's dilemma, we already discussed.

Well, if each of the oligopolists cooperates in holding down output, then high monopoly profits are possible. Each oligopolist, however must worry that while it is holding down output, other firms are taking advantages of the high price by raising output and earning higher profits.

A Prisoner's Dilemma for oligopolist: example

it shows the prisoner's dilemma for a two-firm oligopoly known as a duopoly.

If firm A and B both agree to hold down output, they are acting together as a monopoly and will each earn $1,000 in profits.

However, both firm's dominant strategy is to increase output, in which case each will earn $400 in profits.

Can the two firms trust each other?

Consider the situation of firm A:

- If A thinks that B will cheat on their agreement and increase output, then An increase output too, because for A the profit of $400 when both firms increase output (the bottom right hand choice in (Table X) is batter a profit of any $200 If A keeps output low and B raises output (the upper right- hand choice in the table)).
- If A thinks that B will cooperate by holding down output, then A may size the opportunity to earn higher profits by raising output. Afterall, if B is going to hold down output, then A can earn $1500 in profits by expanding output (the bottom lest hand choice in the table) compared with only $1,000 by holding down output as well (the upper left hand choice in the table).

Thus, firm A will reason that it makes sense to expand output If B holds down output and that is also makes sense to expand output if B raises output. Again B faces a parallel set of decisions.

The result of this prisoner's dilemma is often that even though A and B could make the highest combined profits by cooperating in producing a lower level of output and acting like monopolist,

The two Firms may well end up in a situation where they each increase output and earn only $400 each in profits. So, main concern here now is, how to avoid prisoner's dilemma or how to enforce cooperation.

Individual can use different formal approaches to modify the incentives that decision-makers encounter. Strategies like combined efforts. For enforcing cooperative measures. Through laws, democratic decision making, rules, and precise punitive actions for defections might help in changing numerous prisoner's dilemma into beneficial outcomes.

So, the way out of a prisoner's dilemma is to find a way to penalize those who do not cooperate. A beneficial outcome can happen because cooperation produced better results than defection.

Many real-world oligopolies, provide by economic changes, legal and political pressures, the egos of their top executive, go through episodes of cooperation and competition. If oligopolies could sustain cooperation with each other on output and pricing, they could earn profits as if they were a single monopoly. However, each firm in an oligopoly has an incentive to produce more and grab a bigger share of the overall market; when firms start behaving in this way, the market out come in terms of prices and quantity can be similar to that of a highly competitive market.

SECTION- ONE

The Pandemic Is A Prisoner's Dilemma Game

"A classical game theory case, as people are not taking vaccines in the hope that everyone else would be vaccinated and they would be safe"

- Samarjeet Tripathi

Each individual has choices, but the pay Off for each choice depends on choice made by others. This is what called a prisoner's dilemma game – players weigh cooperation against betrayal, often producing a less than optimal outcomes for the common good.

The perceived benefits and costs of vaccination are often expressed as concerns about side effects and safety. If you are on the fence about vaccination, you might decide- noticing lower infection rates as vaccination campaign gain speed that is no longer seems so critical to get job. Some people might play a 'wait- and – sec game', Dr Bauch said, people who choose not to be vaccinated effectively get a free ride, reaping the benefits of reduced virus transmission generated by the people who do opt for vaccination, but the free rides generates a collective threat.

That is prisoner's dilemma, 'Dr Bauch' said when infection levels are low, people feels less at risk, let down their guard, and then infection levels again rise; the lbb and flow between out behaviour and the virus causes the pandemic waves.

Vaccination decisions based purely on self – interest can lead to vaccination coverage that is lower than what is optimal for society overall. The self-interest strategy maximizing individual payoff is called the Nash equilibrium. so vaccination data(decision) can be influenced by equilibrium and serving the common good.

Each individual has choices, but the payoff for each choice depends on choice made by others. That is what's called a prisoner's dilemma – players weigh cooperation against betrayal, often producing a less than optimal outcome for the common good. The pandemic presents an everyday complexity for such choices. As, if everyone followed public health recommendations: they wore masks, socially distanced, washed their hands, followed COVID-19 appropriate behaviours, stay at home order. In that case there is significantly reduced risk of infection.

But there are always trade-offs and temptations to defect from the regimen. Masks are annoying. Hand washing is tedious. The pandemic is a prisoner's dilemma game played out repeatedly.

Vaccination strategy based purely on self-interest, however game theory assumes people are rational in their decision making. But fear can supress vaccination to precarious levels insufficient to present the spread of an outbreak.

Vaccine hesitancy could be explained by a mathy mechanism called " hysteresis". In general terms, hysteresis occurs when the effects of force persist even after the force is removed. Thus, it is an event that persists into the future, even after the factors that led to that event have been removed. e.g. unemployment rates can remain high even in recovery economy.

Similarly, even a vaccine is seemed safe and efficacious, uptake rates often remain low. The hysteresis effect makes the population hysterical, or sensitive, to the perceived risks of the vaccine.

It boils down to a fundamental problem known as the tragedy of the commons i.e., it occurs when individuals neglect the wellbeing of society in pursuit of personal gain.

Thus, there is a misalignment of individual interests and social interests, to overcome the hysteresis effect, vaccination should be promoted as an of altruism – one's personal contribution of defeating the pandemic.

So, using game theory, researches modelled two ways of prioritising vaccinations i.e., direct and indirect protection, to see which saved more lives and promote vaccination drive and follow all Covid -19 appropriate behaviour necessary urgently to reduce the risk of infection.

SECTION- TWO

Economics

A classic example of the prisoner's dilemma in the real world is encountered when two competitors are battling it out in the marketplace. Often, many sectors of the economy have two main rivals. In the US for example, there is a fierce rivalry between Coca-Cola (ko) and pepsico (PEP) in soft drinks and Home depot (HD) versus lowe's (Low) in building supplies.

Other fierce rivalries include (in India); Reliance versus HP & Jio Versus Airtel and Apple Versus Samsung in the global mobile phone sector. The U.S debt deadlock between the Democrats and Republicans that springs up from time to time a classic example of a prisoner's dilemma.

Let's say the utility or benefit of resolving the US debt issue would be electoral gains for the parties in the next election. cooperation in this instance refers the willingness of both parties to work to maintain the status 940 with regard to the spiralling US budget deficit. defecting implies backing away from this implicit agreement and taking the steps required to bring the deficit under control. If both parties cooperate and keep the economy running smoothly, some electoral gains are assured. but if party A tries to resolve the debt issue in a proactive manner, while party B does not cooperate, this recalcitrance may cost B votes in the next election, which may go to A.

However, if both parties back away from cooperation and play hardball in an attempt to resolve the debt issue, the consequent economic turmoil (sliding markets, possible credit downgrade and government shutdown) may result in, lower electoral gains for both parties.

Advertising is sometimes cited as a real- life example of the prisoner's dilemma. even in anti-trust, authorities want potential cartel members to mutually defect, ensuring the lowest possible prices for consumer.

So, in economics, the prisoner's dilemma can be used to aid decision-making in a number of areas in one's personal life, such as buying a car, salary negotiations and so on.

You must Heard about subsidy?

It may be in form of food, LPG or kerosene. How subsidy case relates to the prisoner`s dilemma? Let`s brainstorm together!

Politicians have used subsidies to entice business, agriculture productivity via fertilizer & seed subsidy, food security and sure renewal energy promotional subsidy e.g. solar panel etc to move their religion since 1930s. but this has led to state and local governments becoming trapped by a prisoner`s dilemma of their own making. The prisoner`s dilemma is perhaps the best known lesson of game theory. The prisoner`s dilemma is a good representative of the situation in which government officials find themselves when businesses solicit subsidies to relocate or expand in a jurisdiction.

In the case of targeted economic development subsidies (TEDS), the site location consultants and businesses act as clever detective by playing local governments against each other, revealing only limited information to extract more of what they want. Government officials face an unenviable choice as the fundamental driving force behind this is a public that widely believes that subsidies do indeed attract business or food security (public distribution system) and financial inclusion (e.g. JAM policy in India). Politicians often justify their subsidy offers by saying they need to cultivate a friendly "business climate", but evidence shows that TEDS are correlated with a higher-tax environment that would likely be less appealing to businesses. A better approach would be to use the subsidy funds to reduce taxes and/or focus on the provision of pure public goods to make the city or state more appealing to all businesses. Industrialist, economic policy makers and politicians know intuitively that`s prisoner`s dilemma.

In my upcoming book on GAEORY; I am analysing subsidy case with prisoner`s dilemma while analysing the famous remark in 1985 by the then prime minister of India that, "Only 15 paise reaches the needy, has found mention in the judgement of the supreme court which said this malaise can be taken care of by Aadhar scheme." so basically, when we analysing both cases comparatively, the dilemma could be solve by JAM Yojna.

SECTION- THREE

International Polities - Partition Plan For Palestine 1947

"War is noble and heroic if a nation is defending it`s sovereignty from the invaders. War is disgraceful and unheroic, a despicable crime against humanity, if a dominant nation is invading another country and killing its innocent men, women, and children. The war, in the 21st century, is ostensibly the methodic game theory in offensive action that has no ethical and moral ground, whatsoever, but to intentionally kill and divide people, and, consequently, paralyze the political and economic stability of another nation."

-Danny Castillones Sillada, The Existentialist Homo Technologicus

In political science & international politics, for instance, the prisoner's dilemma scenario is often used to illustrate the problem of two states engaged in an arm race. both will reason that they have two options, either state will benefit from military expansion. the paradox is that both states are acting rationally, but producing an apparently irrational result. this could be considered as a corollary to deterrence theory.

There are few case Study related to prisoner's dilemma & political science:

Brexit and the prisoner's Dilemma

As of June 24, 2016, Britain and the European union face a real-world prisoner's dilemma. Britain decision to leave the EV by invoking Article 50 means that the country has a two-year period to negotiate its economic and political relationship with the EV. as is expected, both sides are keeping self-interests in mind when approaching the negotiating table.

For example, Britain seeks to obtain the advantages of the EU's single market with non of the potential costs. meanwhile, the EU has called for consequences in regards Britain's exist vote, wanting to make an example for any other country thinking of leaving. however, if both sides decide to work together to come to an Agreements that is mutually beneficial, the prisoner's dilemma can be circumvented and both sides can prosper. If both sides hold fast to personal interests, neither side benefits, it becomes a zero-sum game.

Let's take current burning issue namely Afghanistan's Taliban problem-

Overcoming the prisoner's dilemma to reach peace in Afghanistan:

"If you area leader or someone who works for the interest of a community, first make sure that you understand the interest of the people who make up that community. In this way, you will have good chance of minimizing, perhaps, avoiding the us versus them mentality."

-Duop Chak Wuol

The prisoner's dilemma is a fundamental example in game theory where rational actors (in this case, the united states, the Afghan government and polity, the Taliban, Pakistan and regional actors) fail to cooperate even if they would benefit from doing so.

In this example, each actor would benefit from a peaceful outcome in Afghanistan but many perceive unacceptable short-term costs associated with that outcome. peace in Afghanistan would allow the united states to end its longest war-nearly 20 years of war (recently announces to withdraws its troops) on which it spends approximately $2 trillion.

Afghan government could better invest the billions of dollars op international security assistance it receives in infrastructure; commerce; and the Afghan people, who are among the poorest in the world.

Taliban member could contest for and hold power as a legitimate part of the Afghan polity without combatting international military efforts and sanctions.

Pakistan, which receives widespread blame for providing sanctuary to Taliban senior leaders, could see an improvement in bi-lateral ties with the USA and greater stability on its western border. thus, it is possible to overcome the prisoner's dilemma in Afghanistan and arrive at a mutually beneficial outcome.

(in news: Afghanistan's Ghani government(regime) collapsed/Taliban takes over Afghanistan).

The Two state solution: Israel and Palestine

This two-state, proposed framework for resolving the Israel-Palestine conflict by establishing two states per two peoples: Israel for the Jewish people and Palestine for the Palestinian people.

As it's one of the most protracted of the contemporary period. if cooperation could possible, Jewish get Israel and Arab-Muslim Palestine, even for regional stability. but prisoner's dilemma nature shows because of non-cooperation attitude, the two-state solution unable to resolve conflicts.

SECTION- FOUR

Environmental Studies

According to prisoner`s dilemma, rational individuals might not cooperate even though it would be in their combined best interests to do so. The character of international environmental politics has often been said to resemble a prisoner`s dilemma, same with climate negotiations. (Sooros,1994). This model is applies to different scenarios like the impasse over the Kyoto protocol seemed to resemble a static PD at the moment. Michael Leibreich seeing not only the Kyoto protocol, but International environmental agreements (IEAs) in general, as repeated PDs. Liebreich suggests that the outcome can change since, in the real world, countries can re-assess whether to reduce emissions or not in subsequent negotiations and agreements. So accordingly, the best strategy in repeated Prisoner`s dilemma (PD) would be a "tit-for-tat" strategy:

- Players start by cooperation
- If player defects, then he should be punished in subsequent games until cooperation is reinstated
- Cooperation is reinstated

Think it– how cooperation emerges and persist in environmental agreements and conventions e.g. climate agreements like Paris climate agreement and goals like MDGs (millennium development goals) and SDGs (sustainable development goals-2030).

In Environmental studies, the prisoner's dilemma is evident in crises such as global climate change. all countries will benefit from a stable climate, but any single country is often hesitancy to curb CO2 emissions. the benefit to an individual country to maintain current behaviour is greater than the benefit to all countries if behaviour was changed, therefore explaining the current impasse concerning climate.

In programme management and technology development, the prisoner's dilemma applies to the relationship between the customer and the developer. Capt. Van Ward, an officer in the US Air Force, examined the program manager's dilemma in article published in Defence AT&L, a defence technology journal.

Hello dear! *Are you able to decode; what prisoner`s dilemma encoded?*

Very good. You decoded well. In real world cooperation is more likely when there is room for communication, and likelihood of communication increases in subsequent game. A tit-for-tat strategy also presupposes that players take into account each other`s previous decisions. Cooperative behaviour is more likely in repeated games if players have long-term perspectives and also a mechanism for punishing defection. (Osborne & Rubinstein,1994)

SECTION- FIVE

Animals / Biology

"Paradoxically, it has turned out that game theory is more readily applied to biology than to the field of economic behaviour for which it was originally designed"

-John Maynard Smith

If natural selection is survival of the Fittest, why isn't everything a competition? cooperation and competition have a fitness face off in game theory.

Evolutionary ecologist aims to understand the complex behavioural relationships between organisms as they interact to obtain resources. in general, these interaction strategies, such as combative or cooperative, result in different payoffs based on nature of the interaction. evolutionary ecologists treat these strategies as phenotypes. the most successful organism maximizes their pay off and increase their ability to reproduce. So, biologists utilize game theory to elucidate evolutionary consequences of interactions.

well, cooperative behaviour of many animals can be understood as an example of the prisoner's dilemma. often animals engage in long-term partnership, which can be more specifically modelled as iterated prisoner's dilemma. for example, guppies inspect predators cooperatively in groups, and they are thought to punish non-cooperative inspectors. vampire bats are social animals that engage in reciprocal food exchange. applying the payoffs, from the prisoner's dilemma can help explain this behaviour.

Cooperate/cooperate: Reward: I get blood on my unlucky nights, which saves me from starving. I have to give blood on my lucky nights, which doesn't cost me too much.

Defect/Cooperate: Temptation - you save my life on my poor night. but then i get the added benefit of not having to pay the slight cost of feeding you on my good night.

Cooperate/Defect: Sucker's payoff -I pay the cost of saving your life on my good night. but on my bad night you don't feed me and I run a real risk of starving to death.

Defect/Defect: punishment -I don't have to pay the slight costs of feeding you on my good nights. but I run a real risk of starving on my poor nights.

Other example is the game of chicken (the hawk-dove game) or snowdrift game, the principle of the game used in animal behaviour. it refers to a situation in which there is a competition for shared resource and contestants can choose either conciliation or conflict.

Thus, the application or prisoner's dilemma in even biological application and animal behaviour plays pragmatic role.

SECTION- SIX

What's best for the individual and what's best for society are often not the same thing--this predicament is the premise for the famous "prisoner's

***dilemma" game*". In sport it`s application is too strategic.**

Sport

Doping in professional sports: doping has unpleasant side effects but makes you more likely to win. if everyone dopes, then they're back where they started in terms of the chance they'll win but they've also got vast side effects to deal with. It makes sense to outlaw doping so that all the athletes in a given sport will cooperate in this dilemma.

At its core, game theory is really just about formalizing if/then statements.

In football: If a defence lines up with 7 players to the right of the ball and 4 players to the left side of the ball, then game theory would indicate that the offense's **dominant strategy**is to call a play (any play) toward the left side of the field, taking advantage of the defensive alignment.

In baseball: If historical data suggests that a certain batter has a higher tendency to strike out when a fastball is thrown in the lower right quadrant of the strike zone, then game theory would indicate that the **best response**of the pitcher is to throw a fastball in the lower right quadrant.

(Technical analysis of this section in upcoming book GAEORY)

SECTION- SEVEN

Multiplayer dilemma

The Prisoner's Dilemma also works as a simplified model for competitive interactions between animals, including humans, in the Darwinian struggle for survival of the fittest. If you benefit by either co-operating or defecting in competitive situations where there is a real struggle to survive, then doing so means there is an improvement in your chances of survival and reproduction. So in these basic terms, your chances of survival and reproduction, and so continuation of your genes, can be attributed to your propensity to either co-operate and pass on co-operative genes, or to be ruthlessly selfish (that is, defect) so that the next generation, with similar nasty genes, will also have a higher tendency to be nasty defectors. If defection is always the best strategy in situations of comparative advantage, then all animals, including humans, should have evolved into non-co-operative nasty beings.

Many real-life dilemmas involved in multiple players. although metaphorical, Hardin's tragedy of commons may be viewed as an example of a multi-player generalization of the prisoner's dilemma.

Each villager makes a choice for personal gain or restraint, the collective reward for unanimous (or even frequent) defection is very low pay OFFs (representing the destruction of the commons), such multi-player prisoner's dilemma is not formal as they can always be decomposed into a set of classical two player games. the commons are not always exploited.

William poundston, in a book about the prisoner's dilemma, describes a solution in New Zealand where newspaper boxes are left unlocked.

It is possible for people to take a paper without paying (defecting) but very few do, feeling that if they do not pay then neither will others, destroying the system.

SECTION- EIGHT

Psychology

The Prisoner's dilemma is the standard example of a game analysed in game theory that shows why two completely rational individuals might not cooperate, even if it appears that it is in their best interest to do so.

In addiction research/behavioural economics, George Ainslie points out that addiction can be cast as an intertemporal prisoner's dilemma problem between the present and future selves of the addict. in this case, defecting means relapsing, and it's easy to see that non defecting both today and in the future is by far the best outcome. the case where one abstains today but relapses in the future is the worst outcome-in some sense the discipline and self-sacrifice involved in abstaining today have been "wasted" because the future relapse means that the addict is might back where they started and will have to start over.

The Hunter's Share

This game is a little more life-like than the Prisoner's Dilemma, as it involves more than two protagonists. The aim is to gain the most points, or 'food', after ten rounds or so. Imagine you are one of three hunters. In each round of the game only one succeeds in catching prey, and this is determined randomly by rolling dice, so that for each round it's not predictable who catches it. The amount on the dice also determines the size of the catch – so that a 'two' is a poor catch, and a 'three' is only half of a 'six', and so on. The catch is then shared out. You the hunter can keep all of it for yourself, giving no points for the others; or share it out any way you like, either equally or unequally between the competitors.

So, its application is wide and vast even in ethical learning. Please, do brainstorm work accordingly and prepare case study from different sources like government decisions, judgements and administrative decisions including Administrative Reform Commission(ARC REPORT).

UNIT - III

DISCRIMINE AESTIMATIONELUDI DOCTRINA

"Game theory has an essential role in the analysis of rationality"

- Samarjeet Tripathi

"In game theory, as in applications of the other technologies that use RPT[Revealed preference theory], the purpose of the machinery is to tell us what happens when patterns of behaviour instantiate some particular strategic vector, payoff matrix, and distribution of information – for example, a PD [prisoner`s dilemma]- that we`re empirically motivated to regard as a correct model of a target situation. The motivational history that produced this vector in a given case is irrelevant to which game is instantiated, or to the location of its equilibrium or equilibria. As Binmore (1994,pp.95-256) emphasizes at length ,if, in the case of any putative PD, there is any available story that would rationalize cooperation by either player, then it follows as a matter of logic that the modeler has assigned at least on of them the wrong utility function (or has mistakenly assumed perfect information, or has failed to detect a commitment action) and so made a mistake in taking their game as an instance of the (one shot) PD. Perhaps she has not observed enough of their behaviour to have inferred an accurate model of the agents they instantiate. The game theorist`s solution algorithms, in themselves, are not empirical hypotheses about anything. Application of them will be only as good, for purposes of either normative strategic advice or empirical explanation, as the empirical model of the players constructed from the intentional stance is accurate. It is much-cited fact from the experimental economics literature that when people are brought into laboratories and set into situations contrived to induce PDs, substantial numbers cooperate. What follows from this, by proper use of RPT, not in discredit of it, is that the experimental setup has failed to induce a PD after all. The players' behaviors indicates that their preferences have been misrepresented in the specification of their game as a PD. A game is mathematical representation of a situation, and the operation of solving a game is an exercise in deductive reasoning. Like any deductive argument, it adds no new empirical information not already contained in the premises.

However, it can be of explanatory value in revealing structural relations among facts that we otherwise might not have notices."

CHAPTER FIVE

CHAPTER 5

Appraisal of Game Theory

This chapter includes –

- **Critical appraisal of game theory**
- **Limitation of game theory**
- **importance of game theory**

SECTION- ONE

In this section, I briefly discussed the critical appraisal of game theory.

Critical appraisal of game theory

In terms of game theory, the opposite of a failed tactics is not necessarily success.

-Robert Ferrigno

Criticisms have been levied against game theory, and two in particular are worth nothing.

First, game theory is sometimes judged to have failed, because some games possess solutions and other do not, and those that do have sometimes provide too many.

Here Bacharach (1976) argues that the failure of game theory to give unambiguous solutions to certain classes of games does not imply Flaws or inadequacies in the theory, just that it is the nature of things.

We cannot expect a unique, national answer all the time in a complex, possibly national world.

Second, there is the argument that game theory contours the way individuals view the world, as one of selfish interaction. However, game theory in itself" has no moral content, makes no moral recommendations, is ethically neutral" (Aumann 1987, p.497). It studies selfishness, it does not recommend it.

However, on the Whole the Future of game theory is bright. Recent developments of game theory in economics include the discovery of New Nash equilibrium (Such as the Markoy equilibrium used to study the common resource problem), refinement of equilibrium concepts (such as the stability concept used to study signaling in markets) and evolutionary game theory, an exciting new Field studying social conditioning and behaviour selection (Cho & Krips1987,wcibull1995).

Despite its applicable Functions, game theory is not without criticism, already we discussed. thus, it has been pointed out that game theory can help only so much if you are trying to predict realistic behaviour. Every action, good or bad, can be rationalize in the name of self-interest.

Thus, a constant difficulty with game theory modelling is defining, limiting, isolating or accounting for every set of Factors and variables that influence strategy and outcome.

SECTION-TWO

In this section, I deal with Limitation of game theory.

Limitations of Game Theory

"A thorough understanding of game theory, should dim these greedy hopes. Knowledge of game theory does not make one a better card player, businessman or military strategists."

-Anatol Rapoport

Game theory highlights that, in an oligopolistic market, a Firm behaves strategically, that is, it adopts strategic decision making which means that while taking decisions regarding price, output, advertising etc. It takes into account how it rivals will react to its decisions and assuming them to be rational. it thinks that they will do their best to promote their interests and take this into account while making decisions.

Despite potential Contributions, Game theory has Following limitations:

The biggest issue with game theory is that, like most other economic models, it relies on the assumption that people are rational actors that are self- interested and utility-maximizing. We are social beings who do cooperate and care about the welfare of others, often at our own expense! Game theory cannot account for that fact that in some situations we may fall into Nash equilibrium, and other times we may not, depending on the social context and who the players are.

- Firstly, game theory assumes that each Firms has knowledge of the strategies of the other as against its own Strategies and is able to construct the pay-off matrix for a possible solution. This is highly unrealistic assumption and has little practicability. an entrepreneur is not Fully aware of the strategies available to them, much less those available to his rival. he can only have a guess of his and his rival`s strategies.
- Secondly, the theory of games assumes that both the duopoly are prudent men. each rival moves on this presumption that his opponent will always make a wise move and then he adopts a counter move. this is an unrealistic assumption because entrepreneur is not prudent, he can`t play either the maximin or minimax strategy. Thus, problem can`t be solved.
- Thirdly, various strategies Followed by a rival against the other lead to an endless chain of thought which is highly impracticable

Pay off strategy application with mathematical analysis could be discuss in upcoming book GAEORY, so here just simple analysis done.

there is no end to the thought when A chooses one strategy and B adopts a counter strategy and vice versa.

- Fourthly, it is easy to understand a two-person constant-sum game. But as the analysis is elaborated to three or Four person games, it becomes complex and difficult. However, the theory of games has not been developed for games with more than Four players. for instance, the number of sellers and buyers is quite large in monopolistic competition and the game theory does not provide any solution to it.

- Fifthly, even in its application to duopoly, game theory with its assumption of constant-sum game is unrealistic. for it implies that the

"Stakes of interest" are objectively measurable and transferable. Further the minimax principle which provides a solution to the constant-sum game assumes that each player makes the best of the worst possible situation.

How can the best situation be known if the worst doesn't arise?

Moreover, most entrepreneur act on the presumption of the existence of Favourable market conditions and the question of making the best of the worst does not arise at all.

- sixthly, the use of mixed strategies for making non - zero sum games determinate is unlikely to be found in real market situations. No doubt random choice of strategies introduces secrecy and uncertainty but most entrepreneurs, who like secrecy in business secrets and strategies for the purpose of maximum joint profits.

Thus, like the other duopoly models, game theory fails to provide a satisfactory solution to the duopoly problem. "although game theory has developed for since 1944", writes prof. Watson, its contribution to the theory of oligopoly has been disappointing". to date, there have been no serious attempts to apply game theory to actual market problems, or to economic problems in general.

Despite these limitations, game theory is helpful in providing solutions to some of the complex economic problems even though as a mathematical technique, it is still in its development stage.

SECTION- THREE

In this section, I briefly shed light on the importance of game theory.

Importance of Games Theory

Early in the 20thcentury, mathematics began to study some relatively simple games and later much more complex and the studies regarding game theory begins. game theory is basically denoting on study of mathematically model of differences and cooperation between intelligent balanced decision makers. As rightly said by, Leon A. Petrosjan,1996 that game theory is based on a scientific metaphor, that idea that not only considers as game but also consider like economic competition, War, elections, can be traded and analysed as we would analysis games.

The importance & merits of game theory can be discussed as below:

- Game theory shows the importance to duopolistic of Finding some way to agree. it helps to explain why duopoly prices tend to be administered in a right way. if prices were to change often, tacit agreements would not be Found and would be difficult to enforce.
- Game theory also highlights the importance of self-interest in the business world.in game theory, self-interest is routed through the mechanism of economic competition to bring the system to the saddle point. this shows the existence of the perfectly competitive market.
- Game theory tries to explain how the duopoly problem cannot be determined. For this, it uses the solution without saddle point under constant-sum-two-person game.at the same time, the duopoly problem without a saddle- point is solved by allowing each firm to adopt mixed strategies on a probability basis.in this way, the duopoly problem is shown to be always determined.
- Further, game theory has been used to explain the market equilibrium when more than two firms are involved. the solutions lies in either collusion or non-collusion. These are known as cooperative non-constant-sum-game and non-cooperative non-constant-sum game respectively.
- "Prisoner's Dilemma" in game theory points towards collective decision making and the need for cooperation and common rules of road.

A player in game theory may regarded as a single person or an organisation in the real world subject to decision making with a certain number of resources. The strategy in game theory is a complete specification of what a player will do under each circumstance in the playing of the games. For example, the director of a firm might tell his sales staff how he wants an advertising campaign to start and what should they do subsequently in response to various actions of computing firms.

- The importance of the pay-OFF values lies in predicting the outcomes of a series of alternative choice on the part of the player. Thus, a perfect knowledge of the pay-OFF matrix to a player implies perfect predictions of all factors affecting the outcomes of alternative strategies. Moreover, minimax principle shows to the player the next course of action which would minimise the losses if the worst possible situation arose.

- Again, game theory is helpful in solving the problems of business, labour and management. As a matter of fact, a businessman always tries to guess the strategy of his opponents so as to implement his plan more effectively.

Similar in the case of management in trying to solve the problem of labour union's bargaining for higher wages. Management might adopt the most profitable counter-strategy to tackle such problems. Further, producers might take decisions in which estimation of profits were to be balanced against the cost of production.

Last but not least, there are certain economic problems which involve risks and technical relations. They can be handled with the help of mathematical theory of games. Problems of linear programming and activity analysis can provide the main basis for economic application of the theory of games.

CHAPTER SIX

CHAPTER 6

This chapter includes important concepts related to game theory. For understanding purpose, I use comparisons, tit bits and case study.

This part, I would do detail analysis on my next book on GAEORY.

TOPIC INCLUDES

- Briefly compare between Nash equilibrium and dominant strategy
- Case studies (I would discuss it in my upcoming book GAEORY)

SECTION-ONE

This section deals, the concept of dominant strategy and Nash equilibrium. It compares both while making differences.

Comparison Between Dominant Strategy and Nash Equilibrium

The concept of a Nash equilibrium n-tuple is perhaps the most important idea in noncooperative game theory.

-P. Ordeshook

It is important to compare Nash equilibrium and equilibrium reaches where each firm has a dominant strategy. Whereas a dominant strategy equilibrium describes an optimal or best choice regardless of what strategy the other player adopts, In Nash equilibrium each player adopts a strategy that is the best or optimal, given the strategy other player adopts. However, it may be noted that in some games we do not have Nash equilibrium and that some have more than one Nash equilibrium.

Every dominant strategy is a Nash equilibrium but every Nash Equilibrium is not a dominant strategy.

Suppose, there are 2 players.

In Nash equilibrium, at equilibrium, the strategy of player 2 is the best response to the given strategy of Player 1.

Stated otherwise, no unilateral deviation is profitable for any player.

In Dominant strategy; at equilibrium, the strategy of Player 2 is the best strategy for him whatever the Player 1 chooses.

Same is applicable for Player 1.

Then how every dominant strategy is a Nash?

Nash Equilibrium v. Dominant-strategy Equilibrium Every dominant strategy equilibrium is also a Nash equilibrium, but the reverse is not true. because in every dominant strategy cash, no unilateral deviation is profitable for any player.

Nash Equilibrium v. Rationalizability If a strategy is played in a Nash equilibrium, then it is rationalizable, but there may be rationalizable strategies that are not played in any Nash equilibrium.

So, according to game theory, the dominant strategy is **the optimal move for an individual**regardless of how other players act. A Nash equilibrium describes the optimal state of the game where both players make optimal moves but now consider the moves of their opponent.

UNIT IV

LUDUM DOCTRINA - UT SUBIECTUM: GAEORY

"GAEORY is the subject of next century"

- Samar jeet tripathi

In this unit, I briefly described a new dimension of game theory. Though not elaborated because my main technical & non-technical analysis on GAEORY nonetheless in progress, about to publish next year (2022) on Nobel laureate John Nash`s birth anniversary. The main focus of my analysis is to establish GAEORY as an independent subject. Because as per my analogy, GAEORY is the subject of next century.

CHAPTER SEVEN

CHAPTER 7

"Hey, I am your guest from 22ndcentury"

HELLO, H. SAPIENCE

MYSELF GAEORY, THE SUBJECT OF NEXT CENTURY

Game Theory is the systematic study of strategic interactions that are present everywhere, not only in economics but in politics, sociology, law, biology, sports etc. Thus, game theory applications are omnipresent as can be found in all walks of life.

"Game theory can be seen everywhere in living systems, in general, and human society, in particular. in personal as well as professional life, every day we are faced with decisions which often can be simplified using game theory. There are different areas where game theory has been applied as such (Shoham and Leyton – Brow 2008)."

Game theory should be treated as an independent discipline, though it's in still evolutionary stage.

New independent disciplines are always emerging but only some gain a place in the mother or sister subject.

In fact, most of the now accepted conventional disciplines have, at some stage, struggled to become established.

In the emergence of new subjects, Layton's model outlines a three-stage process for the internal evolution of a new subject.

The evolution of public administration as a discipline can be traced back from political science, as we say, "The scope of the discipline of public administration is determined by what an administrative do" i.e., its scope is boundaryless. but in later evolutionary phase public administration came up with definite scope to be treated as independent subject, MINNOWBROOK conferences infused newness in the discipline:

As per my preliminary research analysis, found game theory potential applications almost everywhere, it's an evolutionary stage to become independent subject, like once public administration was in crawling stage.

So, Game theory evolutionary steps towards as a discipline could be named as GAEORY i.e., the science of Game Theory. In the 21st century, the application of GAEORY is wide ranging. It can be broadly categorised as an activity and discipline.

As a concept, it is new; but game theory (GAEORY) can be traced back to Babylonian/Harappan times. GAEORY as a discipline has been struggling to be a separate subject and not just part of economics and mathematics.

Evolutionary stages of GAEORY with widening scope shows promising step to becoming an independent subject and THE SUBJECT OF NEXT CENTURY.

Evolution takes time, but I see the future of GAEORY; as a discipline and subject of next generation.

GAEORY with its multidisciplinary nature, fulfil the need of every disciplines with a promising future.

Indeed, economics and business have direct impact on our everyday life; so researches treated it as theory of mathematical economics but GAEORY has its scope, action part and philosophical approach.

As a game theorist, my cognitive analysis decodes, that, "GAEORY will be universal subject, theory of generation and discipline of next century".

Our world is undergoing radical, disruptive changes at an accelerating speed – so can we.

We need new discipline for new century so we could sustain the overall development inclusively and manifold. We must not forget; we are in a new era in which information communication technologies (ICT) have become more mature and we are in a new era in which ICT research has shifted its focus from technology development to novel applications.

GAEORY is a theory of generation as per my preliminary research, mainly because of its multi-disciplinary applications. as we see, game theory is everywhere, and you do not have to be an economist to understand its most insightful aspects.

As, GAEORY is a decision marking part so can be seen everywhere, be it economics, business or everyday life etc.

The GAEORY (the science of game theory) is unusual in the breadth of its potential applications, unlike physics, chemistry; which have a clearly defined and narrow scope, the precepts of game theory are useful in a whole range of activities, from everyday social interactions and sports to business and economics, politics, law, diplomacy and war.

Biologists have recognized that the Darwinian struggle for survival involves strategic interactions, and modern evolutionary theory has close links with game theory.

the stuff of Greek methodology; according to Aristotle, the best tragedies are conflicts between a hero and his destiny. They contain reversals of fortune, moments of reorganisation, and ultimately a catharsis.

Game theory is a vast and unique field, game theory is very promising and interesting field of study. It has full potentiality to become as an independent subject, namely GAEORY.

As GAEORY can tell us how to act smart in competitive situations and also can help in our decision making. Thus, indirectly the concepts of game theory are applicable everywhere in our lives.

Little doubt remains that game theory is one of the most important discoveries of modern times and we are yet to understand the full implication of it. so, it has very promising future as GAEORY an independent subject.

In the real world, against or entities are in continuous state of interactions (Niazi). the interactions of agents lead to a wide variety of complexity dynamics (MC Daniel and Driebe2011). so, game theory offers a perspective of analysis of modelling of these interactions (Carmichael, 2005). It's a discipline that studies decision making of interactive entities (Dixit and Skeath1999). Thus, strategic thinking is perhaps the most recognized essence of game theory.

The bottom line is that, GAEORY is a quest for organisation effectiveness(E-3).

ΣPO_s = structure - process – behaviour – environmental

E-3 = Effectiveness / economy/ efficiency

Thus, ***"There are theories in GAEORY but there are no theories of GAEORY"***

this statement intends that, GAEORY is founded on the borrowed theories of mathematics, economics and other related disciplines like behaviour science. Though many theories (applications), which have discovered and born in mother and sister discipline of GAEORY but over exclusively evolutionary stage GAEOY has developed (will develop), theories on its own as. **"Thus, there are both theories of GAEORY and theories in GAEORY".**

GAEORY as an activity is as old as human & economic society but as an academic discipline, scope and orientations have been continuously shifting and blooming.

So, locus and focus of GAEORY still evolving with changing ages, the importance of game theory becomes pivotal. A lot of new technologies are developed, ideologies change, pandemic occurred, changing demography, globalised acceleration, vibrant democratic system set up and last but not least climate changes etc give urgent call to analyse GAEORY as an independent & separate subject.

GAEORY is an universal subject, its application has big impact.

SO, undoubtedly; **"GAEORY is subject of next generation".**

Let's dive in with new dimension of game of theory as a subject i.e. GAEORY

Hold on, my final research analysis in GAEORY will be in book form by 2022(Birth anniversary of John Nash).

CHAPTER EIGHT

CHAPTER 8

MY OPEN LETTER TO ...

Hello homo sapience,

my name is GAEORY. I`ve now my own identity. Since time creation of universe, I co-exist with nature. Nature works as per my rule. I being taught by inter-planetary creatures as well.

Better late than never, thanks to homo sapience to find my identity though not complete but not long to hidden now. I am everywhere, like god but, of course not god. I am mother of science as well as art.

The mysteries that were hidden for ages for the creation and destruction of universe, the law of nature & nature`s law. I am the subject of seekers of truth, who value logics and hear subconscious voice. Every Alchemist knows me so they even illiterate as per your parameter, but at the height of success ladder.

Well, game theory is just part of me, now you have keys but no locks ... smart homo - sapience!

Now, I have name; better give my legal identity while registering in Aadhar. Hold on, here Aadhar is key to get citizenship as proof of domicile though. enrol my name as a separate subject so, scholars could have research on me as a GAEORY, to give me final citizenship card that is – accepting me as a subject.

Well, thanks to Research Scholar Samar J. Tripathi for giving me name as GAEORY and envelop me with aspiration as a separate subject and presenting me before common people to introduced myself.

Thanks to taking interest on me. I am very excited to see my evolutionary journey from a part of mathematical economics to an independent & separate subject.

Guest from 22^{nd}century

- GAEORY

INDUCTIVE ANALYSIS FOR GAEORY

"The implication of game theory, which is also the implication of the third image, is however, that the freedom of choice of any state is limited by the actions of the others."

-Kenneth Waltz

Game theory is everywhere and you don't have to be an economist to understand its most insightful aspects.

From the very beginning, we have tried to predict the future. Prediction is a messy business and astrologers are seldom genuine. However, scientifically speaking, prediction is absolutely essential. From weather to strategic interaction, prediction is present everywhere. It is often impossible to predict complex, real-life phenomena using simple mathematical equations.

So, new concept, like Chaos Theory – which studies complex unpredictable phenomena in terms of strange patters called fractals and some universal laws regarding them. So, Chaos theory studies systems which are unpredictable and slight changes in the initial conditions of which can lead drastically different outcomes. With its wide-ranging applications, Chaos theory is indeed a very promising science.

Other concept- Game theory deals with the mathematics of interaction between players in a game. It`s essentially a science of decision making. Life is analogous to a game and thus game theory also has wide ranging applications in economics, biology and beyond competitions has been scientifically studied since Darwin's time.be it the-survival-of-the-fittest concept in evolution and natural section or competition to gain market share in a modern industry, competition is omnipresent. Essentially all it means is that game theory is a science to study and predict interactions between players in a game.

Recent advances in Game theory have succeeded in describing and Pre-describing appropriate strategies in several situations of conflict and the field. Just as mathematics has carved itself a niche in economics-making it seems unthinkable that a modern economist isn't at least familiar with basic mathematical economics- Game theory has also found itself with a definite role to play economic analysis. The main significant of game theory is to formulate the alternative strategy to compete with one another and in the same since it is an essential tool for decision making process according to fluctuations in relevant contents.

Instead of the limited neo-classical analysis that is tried to only one institutions framework- competitive markets- game theory offers a more flexible analysis of economics problems that includes institutions as well (Bacharch,1976). Thus, game theory is helpful in solving the problems of business, labour and management so in all types of markets, be it so perfect or oligopoly.

Prisoners dilemma in game theory points towards collective decision making and the need for cooperation and common rules on road. The game of prisoner`s dilemma is of important relevance in the oligopoly but it occurs in many aspects of the economy.

Well, the major difficulty that oligopolies face is the prisoner`s dilemma that each member faces, which encourages each member to cheat. Government policy can discourage or encourage oligopolistic behaviour and firms in mixed economies often seek govt blessing for ways to emit competition. As prisoner`s dilemma plays vital role specially in oligopoly because in this market there is tension between cooperation and self-interest So, game theory is a vast and unique field of study. It has full potentiality to become an independent subject i.e. GAEORY.

Game theory is very useful, so its main significant is to formulate the alternative strategy to compete with one another and the same sense it is an essential tool for decision making process according to fluctuations in relevant contents. Ending note for this game theory analysis with Mesquita says;

"Game theory makes strong but open transparent and explicit assumption while many other means of making forecasts are purely judgemental and not transparent. It is inspiring because the terms and ideology are comparatively trouble-free than other theories in this segment."

So, this book, Game theory (A non- technical appraisal of game theory with new dimension), progress toward to establish GAEORY as an independent and separate subject.

IT BEGINS THE JOURNEY OF – GAEORY

IT ALL BEGINS AND ENDS IN YOUR MIND. WHAT YOU GIVE POWER TO, HAS POWER OVER YOU, IF YOU ALLOW IT.

CONCLUDING REMARKS:

Technical approaches of game theory

The discipline of game theory was pioneered in the early 20th century by mathematicians Ernst Zermelo (1913) and John von Neumann (1928). The breakthrough came with John von Neuman and Oscar Morgenstern's book, Theory of games and economic behaviour, published in 1944. This was followed by important work by John Nash (1950-51) and Lloyd Shapley (1953).

Game theory had a major influence on the development of several branches of economics (industrial organization, international trade, labor economics, macroeconomics, etc.). Over time the impact of game theory extended to other branches of the social sciences (political science, international relations, philosophy, sociology, anthropology, etc.) as well as to fields outside the social sciences, such as biology, computer science, logic, etc.

In 1994, the Nobel prizein economics was given to three game theorists, ***John Nash, John Harsanyi and Reinhardt Selten***for their theoretical work in game theory which was very influential in economics. At the same time, the US Federal Communications Commission was using game theory to help it design a $7-billion auction of the radio spectrum for personal communication services (naturally, the bidders used game theory too!).

The Nobel prize in economics was awarded to game theorists three more times: in 2006 to Robert Aumann and Thomas Schelling, in 2007 to Leonid Hurwicz, Eric Maskin and Roger Myerson and in 2010 to Lloyd Shapley and Alvin Roth. In 2012, Alvin E Roth and Lloyd S. Shapley were awarded the Nobel Prize in Economics "for the theory of stable allocations and the practice of market design". In 2014, the Nobel went to game theorist Jean Tirole.

Game theory provides a formal language for the representation and analysis of interactive situations, that is, situations where several "entities", called players, take actions that affect each other. The nature of the players varies depending on the context in which the game theoretic language is invoked.

The nobel prize of economic science, 2020was awarded jointly to paul r. milgrom and robert b. wilson "for improvements to auction theory and inventions of new auction formats".well, we know that, auction theory is a branch of game theory that is used to create mathematical models that can project outcomes for auction bidding processes.

<u>Few points need to analyse -</u>

i. The prisoner`s dilemma comes about when both players attain a noncooperative equilibrium. a noncooperative equilibrium point the property that neither player can gain from violating the equilibrium unilaterally. There would be no dilemma if the players were allowed to talk to face-to-face before they played and to sign enforceable agreements. Keeping the properties of low communication (in particular, communication takes place only through seeing the moves played)/No enforceable agreements, modifications have been suggested for "resolving" the dilemma. They are metagames of Nigel Howard (1966), treatment of the repeated prisoner`s dilemma as a game of infinite length and the games of economic survival or social survival of (Shubik,1964).
ii. The iterated prisoner`s problem game and its various solutions serve as an extremely useful starting point to understand the power and limitations of game theory.
iii. Games of economic and social survival: different approach have been adopted by shubik. In a game of social survival, the objectives of individual are modified to take into account both his gain and his concern for survival.
iv. A game is a description of a strategic environment. Informally, the description must specify who is playing, what the rules are, what the outcomes are depending on any set of actions, and how players value the various outcomes.
v. A game has perfect recall if a player never forgets a decision she took in the past, and never forgets any information that she possessed when making a previous decision.
vi. A key concept in game theory is that of a player's strategy. A strategy, or a decision rule, is a complete contingent plan that specifies how a player will act at every information set that she is the decision

maker at, should it be reached during play of the game.

vii. **Kuhn's Theorem**: For finite games with perfect recall, every mixed strategy of a player has an outcome-equivalent behavioural strategy, and conversely, every behavioural strategy has an outcome-equivalent mixed strategy.
viii. ***Let`s prove that, in 2-player games, a pure strategy is never a best response if and only if it is strictly dominated*** - In games with more than

2 players, there may be strategies that are not strictly dominated that are nonetheless never best responses.[18] As before, it is a consequence of "rationality" that a player should not play a strategy that is never a best response. That is, we can delete strategies that are never best responses. You can guess what comes next: by iterating on the knowledge of rationality, we iteratively delete strategies that are never best responses. The set of strategies for a player that survives this iterated deletion of never best responses is called her set of rationalizable strategies.

ix. **Pure Strategy Nash Equilibrium:**A strategy profile is a pure strategy Nash equilibrium if for all i and $\tilde{s} \in S_i$, $u(S_i, s_{-i}) \geq u(\tilde{S}_i, s_{-i})$. In a Nash equilibrium, each player's strategy must be a best response to those strategies of his opponents that are components of the equilibrium. Unlike with our earlier solution concepts (dominance and rationalizability), Nash equilibrium applies to a profile of strategies rather than any individual's strategy. When people say "Nash equilibrium strategy", what they mean is "a strategy that is part of a Nash equilibrium profile." The term equilibrium is used because it connotes that if a player knew that his opponents were playing the prescribed strategies, then she is playing optimally by following her prescribed strategy. In a sense, this is like a "rational expectations" equilibrium, in that in a Nash equilibrium, a player's beliefs about what his opponents will do get confirmed (where the beliefs are precisely the opponents' prescribed strategies). Rationalizability only requires a player play optimally with respect to some "reasonable" conjecture about the opponents' play, where "reasonable" means that the conjectured play of the rivals can also be justified in this way. On the other hand, Nash requires that a player play optimally with respect to what his opponents are actually playing. That is to say, the conjecture she holds about her opponents' play is correct. Thus, Nash equilibrium is not simply a consequence of (common knowledge of) rationality and the structure of the game. Clearly, each player's strategy in a Nash equilibrium profile is rationalizable, but lots of rationalizable profiles are not Nash equilibria.

x. In the Mixed strategy Nash Equilibrium, each player who is playing a mixed strategy is indifferent amongst the set of pure strategies he is mixing over. The critical need to allow for mixed strategies is that in finite games, the pure strategy space is not convex, but allowing players to mix over their pure strategies "convexifies" the space. This does not mean that mixed strategies are not important in infinite games when the pure strategy space is convex. In many cases, it is claimed that the only way to get around the pure strategy existence problem in the game is through "tricks" such as discretizing the price space, resolving price ties in favour of the lower cost firm (rather than them splitting the market), etc. These do work; they are just not necessary. To sum up, if we allow for mixed strategies, we can always find Nash equilibria in finite games. In infinite games, Nash equilibria may not always exist, but few case gives a broad range of sufficient conditions to assure their existence.

xi. Every finite game has a Trembling handling perfect equilibrium

xii. **Zermelo theorem**: Every finite game of perfect information has a pure strategy Nash equilibrium that can be derived through backward induction. Moreover, if no player has the same payoffs at any two terminal nodes, then backward induction results in a unique Nash equilibrium.

xiii. **Auction theory** can be applied to a vast amount of fields and situations, but is perhaps most well known for being used in economics. GAEORY will discuss this theory with real life applications.

xiv. a **Bayesian game**is a game in which players have incomplete information about the other players. For example, a player may not know the exact payoff functions of the other players, but instead have beliefs about these payoff functions. These beliefs are represented by a probability distribution over the possible payoff functions.

References

1. Nash, J. (1950a). 'Equilibrium Points in *n*-Person Games.' *Proceedings of the National Academy of Science*/(1950b). 'The Bargaining Problem' *Econometrica* /(1951). 'Non-cooperative Games.' *Annals of Mathematics Journal*, / (1953). Two-Person Cooperative Games. *Econometrica*,
2. Sigmund, K. (1993). *Games of Life*
3. Ross, D., and LaCasse, C. (1995). 'Towards a New Philosophy of Positive Economics'. *Dialogue*
4. Schelling, T (1960). *Strategy of Conflict*
5. *R. Axelord, "The evolution of cooperation" (1984)*
6. *Hal R. Varian, Intermediate Microeconomics : A Modern Approach*
7. *"Rational Cooperation in the finitely Repeated Prisoners` Dilemma, journal of economic theory; work done by Davis Kreps, Paul Milgrom, John Roberts and Roberts Wilson*
8. *J. Tirole, the theory of industrial organisation*
9. *W.J.Baumol, on the theory of oligopoly Economica & Economic theory and operationas analysis*
10. *John Von Neumann and Oscar Morgenstern, Theory of Games and Economic Behaviour*
11. *Brams, Steven, Game theory and politics*
12. *Ariel Rubinstein , Economic Fables*
13. *Abhinay Muthoo; Bargaining theory with applications*
14. *Hans Peter; Game theory : A multi-levelled approach*
15. *Frank C. Zagare; Game theory : concepts and Applications*
16. *Eric Rasmusen; Games and information: An introduction to game*
17. Shelling T. (1980). The Intimate Contest for Self-Command. *Public Interest*
18. Weibull, J. (1995). *Evolutionary Game Theory*
19. Binmore, K. (1994). *Game Theory and the Social Contract* (v. 1): *Playing Fair*/(1998).*Game Theory and the Social Contract* (v. 2): *Just Playing*
20. Double resonance in cooperation induced by noise and network variation for an evolutionary prisoner's dilemma." *New Journal of Physics.*

www.ingramcontent.com/pod-product-compliance
Lightning Source LLC
Chambersburg PA
CBHW021449210526
45463CB00002B/699
9781685863159